S. Itakura, K. Fujita (Eds.)

Origins of the Social Mind

Evolutionary and Developmental Views

S. Itakura, K. Fujita (Eds.)

Origins of the Social Mind

Evolutionary and Developmental Views

Springer

Shoji Itakura, Ph.D.
Associate Professor, Department of Psychology
Graduate School of Letters, Kyoto University
Yoshida-Honmachi, Sakyo-ku, Kyoto 606-8501, Japan

Kazuo Fujita, Ph.D.
Professor, Department of Psychology
Graduate School of Letters, Kyoto University
Yoshida-Honmachi, Sakyo-ku, Kyoto 606-8501, Japan

***Cover*:**
Upper: Chimpanzees. Photo by Satoshi Hirata
Center left: Capuchin monkey. Photo by Kazuo Fujita
Center middle: Keas. Photo by Camelon Mulvey
Center right: Human children. Photo by Yuko Itakura
Lower: Scrub-Jay. Photo by Nathan Emery

ISBN 978-4-431-75178-6 Springer Tokyo Berlin Heidelberg New York
e-ISBN 978-4-431-75179-3

Library of Congress Control Number: 2008928681

Printed on acid-free paper

Springer is a part of Springer Science+Business Media
springer.com

Typesetting: SNP Best-set Typesetter Ltd., Hong Kong
Printing and binding: Hicom, Japan

Preface

During the first few years of life, children acquire knowledge about the relationships between their own mental states, their actions, and the social world. This information is then used to understand themselves and others. Humans are born into families, where they are raised and learn to cooperate, compete, and communicate. We are social creatures, and over the course of development, we learn about people, relationships, social systems, and others' minds. In addition, not only do we live socially, we think socially as well. However, human adults are not the only creatures to live and think socially. In recent years, sophisticated experiments have provided new information about social cognition in human infants and nonhuman animals.

In this book, we focus on the developmental and evolutionary origins of the social mind, bringing together the currently segregated views on social cognition in the two fields. Ever since the term "theory of mind" was coined by D. Premack nearly 30 years ago, the concept has been the main topic of social cognition research both in developmental psychology and in primatology. However, few attempts have been made to integrate these two research domains. Just recently, researchers from the two areas collaborated to publish a book on this topic, but the volume was little more than a collection of independent papers. This book overcomes that limitation by presenting new data and their implications from both developmental and evolutionary points of view. The subjects examined in this book are humans, nonhuman primates, and birds—in particular, corvids and a parrot. We are able to consider the phylogeny of the social mind by comparing humans and nonhuman primates. We are also able to discern the environmental factors (i.e., selective pressures) or the evolutionary processes of the social mind by comparing and contrasting differentiated species. Most important for this examination, it is first-class researchers, active in their own fields, who present and discuss their work.

The book has two parts. The first consists of five chapters dealing with the phylogenetical aspects of social intelligence and social cognition in nonhuman animals. Each contribution provides evidence of social cognition or social intelligence in different species. Fujita et al. (Chap. 1) provide a variety of evidence of social intelligence in capuchin monkeys. In a series of experiments they demonstrate that capuchin monkeys are able to deceive a dominant opponent tactically, cooperate with one another, are sensitive to the subtle states of human eyes, recognize knowledge states of human trainers, and so on. All primate species experience their first social interaction with their parents. Hirata (Chap. 2) reports on communication in mother–infant dyads and proposes the possibility of its role in the evolution of social intelligence in chimpanzees. Anderson and Vick (Chap. 3) conducted an overview of the literature on gaze processing, which is considered very important in primate social behavior. Many species of nonhuman primates reliably and spontaneously demonstrate visual co-orientation in response to changes of both head and eye direction. Then what is the underlying social intelligence in nonhuman primates? Emery and Crayton (Chap. 4) took on the challenge of clarifying what constitute species-specific social cognition. They provide evidence of a theory of mind in scrub jays by analyzing in detail the different cache protection strategies used by western scrub jays. Huber et al. (Chap. 5) demonstrate that the collaborative solution of instrumental tasks depends on two main factors, cognitive competence and favorable social conditions, which are the basis of their work with keas.

The second part of the book consists of four chapters, providing empirical findings or a theoretical framework on the developing ability of human children to negotiate with others, to tell a lie and to understand others' mental state, and of adults' development of the capacity for vicarious experience. Ando and Koyasu (Chap. 6) examined how acting skills develop by comparing actors at three levels of expertise. They find that it is important to act in accordance with what is needed in the scene, taking into account not only the audience but also the setting as a whole. Rochat and Ferreira (Chap. 7) maintain that negotiation is essentially a conversation that with development is increasingly initiated by the child in the form of bargaining. From the perspective of ontogeny they discuss how children become *Homo negotiatus*. Talwer and Lee (Chap. 8) review and evaluate the current findings about the development of verbal deception in terms of the types of lies that children tell, their motivations for lying, age-related changes in lie-telling skills, and the cognitive and social factors that are associated with lying. Itakura et al. (Chap. 9) focus on the development of mentalizing in human children. They raise two questions concerning this issue: 1) When do children first attribute a mental state to others? and 2) When they do so, to whom do they attribute a mental state? They also propose a new research domain to be called "Developmental Cybernetics." Finally, Itakura offers a commentary on the integration of research in developmental science and comparative cognitive science.

Attempts to integrate research from the two domains have been reflected in recent symposiums and workshops all over the world that have brought together comparative psychologists, primatologists, and developmental psychologists. Here, we present the first book that parallels that trend.

Shoji Itakura and Kazuo Fujita

Contents

Part I
Phylogeny of Social Cognition

1
Social Intelligence in Capuchin Monkeys (*Cebus apella*)

Kazuo Fujita[1], Hika Kuroshima[2], Yuko Hattori[3], and Makoto Takahashi[1]

1. Introduction

Asking how socially intelligent capuchin monkeys are seems interesting and important for at least two reasons. The first reason is that they are New World monkeys, and answering this question gives us insights into the origin of this aspect of intelligence in the primate order. This insight is possible because the social intelligence of prosimian species seems quite limited; for instance, there have been no reports of potential cases of deception from the field observations in this group (see Byrne 1995), and triadic interactions seem rare among prosimians despite their apparently social nature (Jolly 1988). Thus, the origin of advanced social intelligence in primates may be traced back at the maximum to the common ancestor between New World monkeys and Old World monkeys that lived about 30–35 million years ago.

The second reason is that capuchin monkeys are the best primate users of tools both in the laboratory and in the wild, if we exclude great apes, and thus seem to have advanced physical intelligence (Fujita et al. 2003; Sato et al., submitted; Visalberghi 1990). Wild capuchins even crack open nuts placed on a hard surface with a hammer stone (Fragaszy et al. 2004) as do chimpanzees. The social, or Machiavellian, intelligence hypothesis (Byrne and Whiten 1988), which stresses complex social life as a most powerful pressure for advanced information-processing ability, predicts good social intelligence in capuchin monkeys. Thus, answering the foregoing question also examines this hypothesis.

Below we report a series of studies conducted at our laboratory in Kyoto University that addressed these questions with captive tufted capuchin monkeys (*Cebus apella*). First, we describe the two studies on the most complex social skills; one is deception and the other is cooperation between monkeys. Second, we describe several tests of various aspects of the more fundamental social recognition underlying such complex social skills.

[1]Graduate School of Letters, Kyoto University, Yoshida-Honmachi, Sakyo-ku, Kyoto 606-8501, Japan
[2]Japan Society for the Promotion of Science; Department of Psychology, University of Georgia, Athens, GA 30602-3013, USA
[3]Japan Society for the Promotion of Science; Primate Research Institute, Kyoto University, Inuyama, Aichi 484-8506, Japan

2. Experimentally Induced Spontaneous Deception in Capuchin Monkeys

Although there have been numerous reports of deception in great apes and Old World monkeys, Byrne (1995) reports only four observations of possible deception in capuchin monkeys. However, this seems odd considering the fact that tufted capuchins have an ability for recognition of human-given cues comparable to or even better than that of rhesus monkeys (Anderson et al. 1995; Itakura 1997; Itakura and Anderson 1996; Vick and Anderson 2000). It is true that in some cases this ability could develop irrespective of the taxon; dogs outperform chimpanzees in object-choice tasks using human-given cues (Hare et al. 2002). Dogs could be a special case because of their prolonged domestication. Given this well-developed aspect of social recognition by capuchin monkeys, we are tempted to test whether they spontaneously start to deceive opponents in a situation where deceptive acts could benefit themselves.

Actually, tufted capuchins learn to behave in two-choice tasks deceptively to competitive human trainers who take food found in containers that the monkeys point at (Mitchell and Anderson 1997). This action may be regarded as a case of deception by this species. However, because the monkeys who succeeded in "deception" were explicitly rewarded by the experimenter after deceptive pointing, obviously the behavior could also be a result of simple operant conditioning and may not incorporate any understanding of the mental aspects of the human trainers.

We devised a situation in which monkeys could spontaneously start to deceive conspecific opponents (Fujita et al. 2002). The situation is a food competition contest. Two participants, one dominant and the other subordinate, faced each other in two transparent cages. Two food boxes were placed between the participants. One side of each food box was transparent and the other side was opaque. The food boxes could be opened only from the transparent side, which faced the subordinate monkeys, by pulling a little handle on the lid. The dominant subject always faced the opaque side of the food boxes (Fig. 1).

Before the monkeys had a food competition contest using this apparatus, we trained them in fundamental skills of how to manage the food boxes. Four subordinate monkeys learned to open the baited one of the two boxes by pulling the handle to take food out at an accuracy of 100%. One dominant monkey learned to take the food in the box by manually inspecting inside immediately after the experimenter opened it.

Each of the subordinates had ten contests with the dominant monkey daily. In each contest, an opaque screen was placed at baiting to occlude the dominant's view and a clear screen to prevent the subordinates' reaching. Each contest started with first removing the opaque screen on the dominant's side and then, after 5 s, removing the clear screen. Thus, only the subordinates were able to see the food and to open the baited box at time of the contest. However, the dominant monkey was able to usurp the food once the box was opened. A possible

FIG. 1. The food competition contest between two capuchin monkeys. *Left:* The subordinate monkey on the right tries to open the baited box placed between the two cages. *Right:* The dominant monkey on the right tries to usurp the food in the opened box. (From Fujita et al. 2002)

strategy of the subordinates was to open the empty box first to attract the attention of the dominant, then to switch to the baited box.

During the course of six to eight sessions, two of the four subordinates started to open the empty box first. This "reverse opening" accounted for more than 10% of the contests these two monkeys participated in. This strategy could be deceptive, but it could also be a mere consequence of decrease in the rate of obtaining the reward because of the dominant's usurping.

To examine this possibility, in the second experiment we modified the food boxes to install an "automatic food dropper," which was actually a set of a hole and a magnetized plate at the bottom of the food box. This mechanism dropped the food down to the floor only when the magnetized plate was connected to the lid of the box by a string.

With a pair of these modified food boxes, we tested the two "deceptive" monkeys individually. We set the rate of this automatic "loss" of food at the same value as that of the previous five sessions of the contests in which the monkeys faced the dominant, who was actually absent in this test. During the four or five test sessions consisting of ten trials, the monkeys showed the reverse opening in just two trials. Thus, the reverse opening shown in the first experiment was not accountable as a consequence of decreased rates of obtaining a reward.

In the third experiment, both subordinates were retested with the dominant monkey using the modified food boxes for five sessions. The reverse opening by one of the subordinates revived to 10% but that by the other monkey disappeared totally. We analyzed the positioning of this latter monkey during the 5 s after removing the opaque screen on the dominant side and before removing the transparent screen on the subordinate side, when the dominant could get prepared for usurping. This post hoc analysis revealed that this monkey had tended to position himself to the nonbaited side. This observation might suggest that he attempted a milder deception toward the dominant monkey who might behave to him in a mean fashion in the home cage where they live together.

The reverse opening by the two monkeys was not actually effective in increasing the rate of obtaining the reward. The monkeys did not always switch to the baited box immediately after opening the empty box. This might suggest that the reverse opening was not planned. However, the subordinates could have been afraid of revenge by the dominant in the home cage as well. It is difficult to conclude whether the subordinates' reverse opening, and reverse positioning, incorporated deceptive intent or not. However, failure of increasing the rate of obtaining the reward on the other hand suggests that these behaviors cannot be explained by simple operant conditioning of such spontaneous novel acts.

3. Spontaneous Solution of a Cooperation Task and Possible Reciprocal Altruism Between Monkeys

Cooperation has been difficult to understand from an evolutionary point of view, because the helper often immediately loses resources such as time, energy, potential mates, etc., by performing the cooperative act. The puzzle could be solved by kin selection, which leads to an ultimate gain of reproducing the genes of the helper shared with the helped kin, or by reciprocal altruism, which benefits the helper in later occasions when the helped individual helps the original helper. Although the former mechanism has been proved in social insects such as bees and ants and in some vertebrates, the latter has not been well documented except in an often-cited case of vampire bats.

In a shorter time scale, cooperation may occur when two individuals must work together to collect otherwise unobtainable rewards. For instance, Mendres and de Waal (2000) demonstrated that two capuchin monkeys in adjacent cages synchronously pulled a heavy plate with foods on it. Pulling decreased when there is no monkey in the next cage or when the monkeys were unable to see the next monkey. Pulling continued when only one of the participants obtained food sharable through the grid. On the other hand, Visalberghi et al. (2000) reported that capuchin monkeys succeeded in pulling the plunger synchronously without reference to the partner's behavior; that is, randomly pulling the plunger was occasionally synchronized between the participants. Thus, this type of cooperation may not warrant understanding of the partner's role, or possibly intention to cooperate or to request the partner to help.

Several previous studies trained separate roles to individuals that would be "pasted" between participants afterward. Chimpanzees are quite excellent in this type of task. For instance, Savage-Rumbaugh et al. (1978) demonstrated that two language-trained chimpanzees cooperate with different roles to obtain rewards packed in the box via symbolic communication. One individual requested a tool to open the baited box using lexigrams to the other in the next room where there were various tools. The second chimpanzee handed the requested tool to the first after reading the facsimiled lexigram. Povinelli et al. (1990) trained chimpanzees to pull the handle that the experimenter pointed at or to inform the experimenter which handle to pull. When the role was reversed, the chimpanzees performed appropriately to the situation. However, the same experiment in rhesus monkeys by Povinelli et al. (1991) failed in such role reversal. Also, cooperation similar to the symbolic communication by chimpanzees was obtained in pigeons, although they failed in role reversal (Epstein et al. 1980). Thus, this "pasted-parts" type of cooperation can also occur as an interindividual chain of responses without understanding the partner's role or intention.

We tested tufted capuchin monkeys in a third type of cooperative problem. We first trained them individually to perform a full sequence of behavior to obtain rewards. Then we tested whether the monkeys were able to spontaneously divide the behavioral sequence between the two (Hattori et al. 2005).

The behavioral sequence was to pull a small tab, which blocked the movement of a long stick, off the pathway attached to the outer wall of the cage, at one end of the experimental cage (box A), then to push the stick along the pathway at the other end (box B). When monkeys completed this sequence by themselves, they were able to collect a piece of food in box A, placed in a den under the stick, and another piece of food that dropped into a food cup from the pathway in box B (Fig. 2). Six monkeys successfully learned to perform this sequence smoothly.

In the cooperation test, the two cages (boxes A and B) were separated by a transparent board and two monkeys were each entered into each box. In this situation the monkeys had to perform each separate action—pulling the tab and pushing the stick—in each box to collect food.

Before going into this cooperation test, we carefully trained each monkey not to touch the experimental panel for 1 min with the transparent partition. The partition was removed after 1 min and the monkey was allowed to conduct a full sequence of actions.

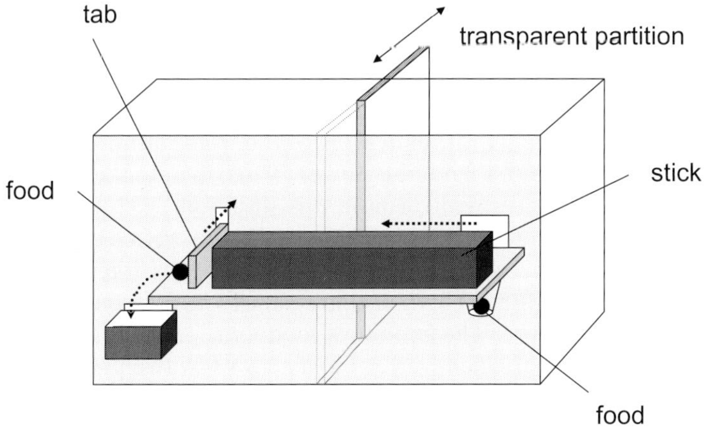

Fig. 2. A schematic drawing of the apparatus to test cooperation between two capuchin monkeys. See text for details. (From Hattori et al. 2005)

The six monkeys were grouped into three non-kin pairs. We tested the monkeys for two sessions of ten trials. Trials ended when the monkeys collected food or after 1 min. One of the three pairs started to cooperate in the second trial of the first session. In the second session in which the roles were reversed, this pair started to cooperate in the third trial. The other two pairs never cooperated in this first series of tests. However, in the second series conducted after retraining them in the original task, both pairs started to cooperate in the first trial of both the first session and the second session in which the roles were reversed. In all successful cases, the time to success in each trial drastically decreased. Thus the monkeys seem to have solved this cooperation task insightfully rather than by trial-and-error.

Next, we tested whether this cooperation was an outcome of a mere interindividual chain of responses or whether it incorporated any communication between monkeys. We devised two situations. One was exactly the same as the test trials in the test above (cooperation condition). The other was similar, but the food in box A was buried in the tab and the stick in box B was short enough to move without pulling the tab off (solo condition). The monkeys' behavior was video analyzed. We found that the monkey in box B looked at the monkey in box A significantly longer in the cooperation condition than in the solo condition. The monkey in box A also tended to look at the monkey in box B longer in the cooperation condition, but this did not reach statistical significance. In the solo condition, the monkeys typically started their own actions apparently without paying attention to their partner next door. In the cooperation condition, in contrast, the monkeys often came up to the partition to check the partner's behaviors. The results reject the simple interindividual chain account and rather suggest cooperation incorporating communication between participants.

Finally, we tested whether this cooperation might advance to reciprocal altruism between monkeys. In this test, there was no food placed in box A. The only food was under the stick. Thus, the monkey in box B needed help by the partner in box A to collect food, but the monkey in box A obtained no food for his/her help. We tested the three pairs of monkeys by exchanging their placement between the two boxes every trial for two or three sessions of ten trials. As a result, all three pairs maintained cooperation in most of the trials, although pulling the tab by the monkeys in box A tended to be slower. Thus, the monkeys showed behavior interpretable as reciprocal altruism in this situation.

In summary of this series of cooperation experiments, tufted capuchin monkeys spontaneously solved the cooperation task in which division of labor between the participants was required. This solution seems to have been a result of insight rather than trial-and-error. The monkeys' cooperation does not seem to be a consequence of accidental synchronization of actions by two participants. Finally, the monkeys demonstrate possible reciprocal altruism when only one of the participants was rewarded but their roles were switched every trial. The last result suggests that capuchins have such intelligence as to pay costs for benefits in the future in social interactions.

4. Underlying Subprocesses

The above-stated experiments suggest that tufted capuchin monkeys have very well developed social intelligence; first, they may take spontaneous acts interpretable as deceiving conspecific opponents in an experimental food competition contest without explicit training to do so. Second, these monkeys spontaneously solve a cooperation task by dividing a series of familiar acts between two participants and, furthermore, they show reciprocal altruism in a modified task in which only one of the participants is rewarded at a time.

However, notwithstanding such careful tests conducted as reported above, it is still possible to argue that the monkeys' behavior may be no more than "blind" operants because the monkeys have been rewarded by such responses. This type of account by learning can never be rejected completely as it can assume various rewards that are not experimentally manipulated, including intrinsic ones, and generalization of the responses to novel situations. We believe such an account to be nonproductive, and any attempt to reject it could be exhausting. Thus, we took another strategy in examining whether the potentially tactical social intelligence incorporates understanding others' mental states (mentalizing; Frith et al. 1991).

There are various cognitive processes seemingly underlying deception and cooperation. Such subprocesses include recognition of other's perspective, attentional states, knowledge, intention, desire, characteristics, causality in social situations, etc. Examining these subprocesses in capuchin monkeys helps us to understand the nature of social intelligence of this species, and their successful demonstration supports the possibility that the deceptive and cooperative behavior shown by capuchins actually incorporates mentalizing. The experiments that follow addressed some of these processes in capuchin monkeys.

4.1. Recognition of Another's Gaze and Attentional States

Previous reports suggest that monkeys have difficulty in understanding a human actor's attentional focus suggested by eyes (Anderson et al. 1995; Itakura 1997). Some reports suggest that even chimpanzees may do. For example, Povinelli and Eddy (1996a,b) reported that chimpanzees have difficulty in differentiating their food request toward two trainers in different attentional states, except between one fully faced to the chimpanzee and the other fully turned away. A better but similar result was obtained by Kaminski et al. (2004). In their experiment, requesting by chimpanzees and other apes was more frequent toward a human sitting with back turned but with the face turned forward than toward a human sitting with the back fully turned. However, the same apes failed to differentiate between a human with eyes open and a human with eyes closed.

Such failure seems odd because infant chimpanzees have been shown to prefer human photos of direct gaze to those of averted gaze and those with eyes open to those with eyes closed (Myowa-Yamakoshi et al. 2003). It is possible that chimpanzees lose sensitivity to the eyes as they grow up, but it is more likely that some procedural difference causes this contradiction. We tested whether tufted capuchin monkeys would recognize attentional states of the experimenter shown by the eyes using two different measures: that is, pointing and looking (Hattori et al. 2007).

Our capuchin monkeys were first trained to reach toward the baited container rather than the empty one in a two-choice situation. The experimenter immediately opened the container and handed the food inside to the monkeys after this pointing gesture. The containers were transparent, thus this task was quite easy for the monkeys to learn. Next, in the test that followed, two test trials were inserted among ten training trials. As before, the experimenter handed the food to the monkeys immediately in those training trials. In the test trials, however, the experimenter ignored the monkeys' reaching responses and acted either of the following two roles: looking at the monkey or looking up at the ceiling. We analyzed the frequency of monkeys' pointing behavior and the duration of looking at the experimenter. The frequency of pointing did not differ between the two actions. However, the monkeys looked at the experimenter longer in the monkey condition than in the ceiling condition (Fig. 3a). This result suggested that the monkeys recognized the direction of the experimenter's eyes, but this discrimination did not come out as differential pointing.

In the next experiment, the experimenter acted the following: to look at the middle of the two containers with eyes open and to "look at" the same location with eyes closed. The frequency of the pointing again did not differ between the two actions. However, the monkeys looked at the experimenter longer in the eyes-open condition than in the eyes-closed condition (Fig. 3b). Therefore, the monkeys recognized the state of the eyes of the experimenter. Again, however, this discrimination failed to show up as differential pointing.

These results demonstrate that tufted capuchin monkeys are sensitive to the subtle state of human eyes, just like infant chimpanzees reported by Myowa-Yamakoshi et al. (2003). However another important finding is that the pointing gesture made by this species does not reflect this discrimination. Such dissociation may suggest that the pointing behavior trained through an artificial task may not be a communicative act but an arbitrary operant. Otherwise, this may suggest that implicit recognition of the attentional states may not warrant explicit recognition that could lead to differential requesting behavior. Such dissociation has been also shown in human infants in object permanence tests (Baillargeon 1993) and false-belief tests (Onishi and Baillargeon 2005).

4.2. Inference of a Consequence of the Other's Behavior

To outwit or cooperate with others, one must infer the consequence of the other's behavior. This aspect of social intelligence by nonhumans has not been well analyzed. We tested whether tufted capuchin monkeys are capable of this type of inference (Takahashi et al., in preparation).

The situation was a feeding context. Four feeding sites were set inside two adjacent rooms (Fig. 4). A small cage having two doors on both ends was placed

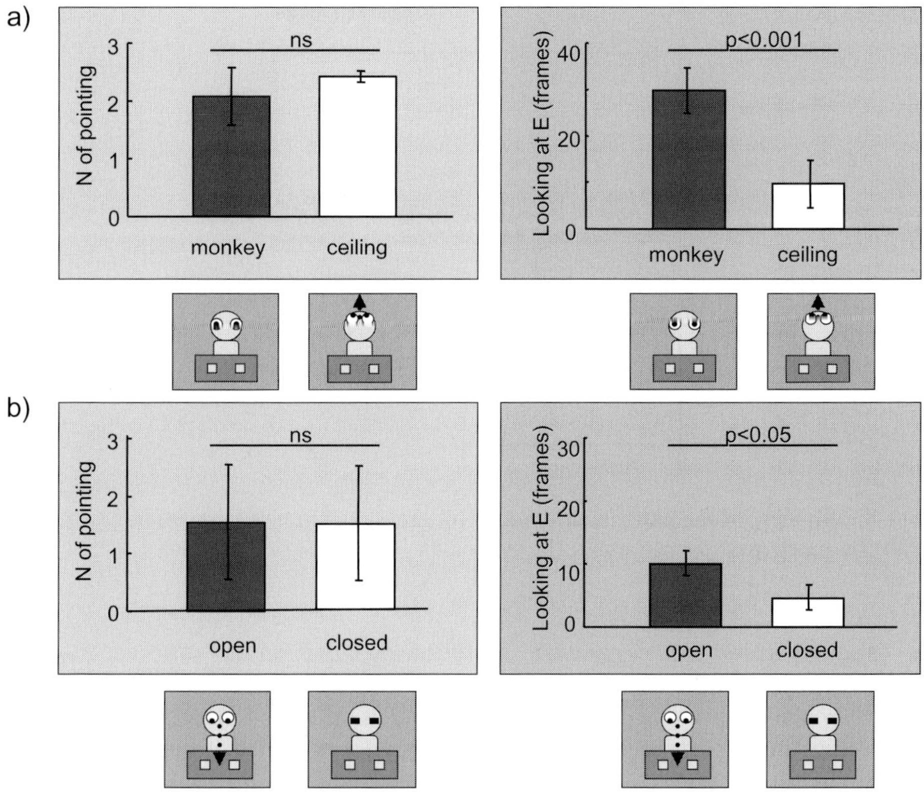

FIG. 3. The number of pointings (*left*) and the duration of time of looking at the experimenter (*right*) while the experimenter showed different acts showing her attentional states. **a** When the experimenter looked at the monkey or looked up at the ceiling. **b** When the experimenter looked at the location between the two containers with her eyes open or with her eyes closed. (From Hattori et al., in preparation)

between the rooms, so that it made two rooms communicating, from which the monkey was able to see the direction of the four feeding sites. However, direct view of the feeding sites could be occluded by opaque screens placed near the sites.

First, four monkeys were trained to forage at these feeding sites. When the inner two sites were used, monkeys were able to eat food each time they visited the sites up to five times, during which the experimenter placed food every time the monkeys ate it and left. Thus, these inner feeding sites were "replenishing" sites. When the outer two sites were used, the monkeys were able to eat food only at their first visit to the sites. Thus, these outer sites were "expiring" sites. The opaque screens signified which sites were used for the trial. Three trials of each condition were run per day, and this training continued for 25 days.

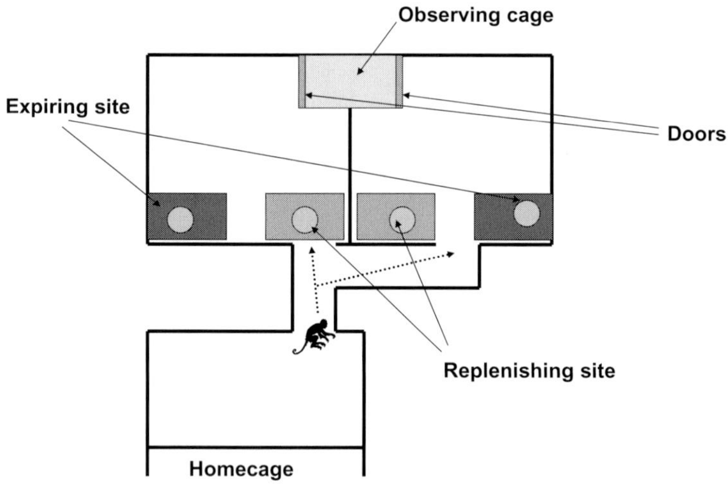

Fig. 4. Top view of the experimental setup to test capuchin monkeys' inference of food location based on the social clues. (From Takahashi et al., in preparation)

Then, we tested the monkeys in the following situation. An observer monkey was first placed in the cage located between the rooms. A demonstrator monkey was then placed into one of the two rooms and ate the food placed in the feeding site. After the demonstrator left the room, the doors of the observer's cage was opened and the observer monkey was allowed to choose one of the sites.

We recorded the first visit of the observer monkey. Our prediction was as follows: If the observer is able to infer the consequence of the demonstrator's behavior, then the observer will visit the site unvisited by the demonstrator first more often when expiring sites are used than when replenishing sites are used.

To reject the possibility of simple learning by experience, we randomly rewarded the observer's visit to the feeding sites at the probability of 0.5. In total, 24 trials were tested for each type of feeding sites. Three of the four capuchin monkeys visited the unvisited sites first in about 70% to 80% of the trials for the expiring sites, which was significantly more often than for replenishing sites. It may be noteworthy that the proportion of visiting unvisited sites was as low as about 20% to 30% of the trials for replenishing sites; this means that these successful monkeys tended to visit the same site that the demonstrator visited when replenishing sites were used. This preference seems to be a natural and adaptive characteristic of this social species; they may like to follow the leader who may know about the resource better.

Therefore, capuchin monkeys are clearly capable of inferring possible consequences of conspecific behavior. We actually tested three nonprimate species, tree shrews (*Tupaia belangeri*), golden hamsters (*Mesocrecetus auratus*), and rats (*Rattus norvegicus*), in basically the same procedure. None showed successful inference. Thus, this type of inference in a social situation might not be widespread among mammals.

4.3. Understanding Knowledge Status in Others

Recognizing what others do and do not know is also essential to deceive or cooperate with others. It is difficult for anyone to deceive others who know the truth. Observational studies of spontaneous deception and counterdeception in great apes suggest they have this ability. For example, Menzell (1974) observed a series of deceptive tactics by young chimpanzees. He taught a female named Belle where food was hidden in the enclosure. Belle was returned to the group cage next to the enclosure, then all individuals in the cage entered into the enclosure. Belle at first led the group to the food site but a dominant male named Rocke eventually started to monopolize the food. Belle came to avoid going to the food site directly but to remain some distance from the goal until Rocke's attention averted. Rocke now started to pretend to be uninterested but actually was monitoring Belle. The whole interaction became more and more complicated as the trial was repeated. Such complicated sequences demonstrate the chimpanzee's knowledge of another's knowledge. Coussi-Korbel (1994) observed a similar interaction in mangabeys, although he suggested a simple operant learning by this species.

In a more controlled situation, Povinelli et al. (1990) showed that chimpanzees recognized that seeing leads to knowing. In this experiment, a human actor, called the "hider," hid food in one of the four opaque containers behind the screen. A second actor, called the "knower," watched the behavior of the hider and thus was knowledgeable of the location of food. A third actor, called the "guesser," left the room before the hider hid food and thus was ignorant of the food location. The guesser returned to the room after a whole sequence was completed. The two latter actors, the knower and the guesser, "advised" the chimpanzees where the food was by touching the containers. The knower always touched the correct container and the guesser touched one of the empty containers. The chimpanzees eventually learned to pull the container the knower touched significantly more often than the one the guesser touched. Four-year-old human children were successful in the same experiment, but 3-year-old children were not (Povinelli and deBlois 1992). Rhesus monkeys showed no signs of learning in the same experiment (Povinelli et al. 1991).

Recognizing knowledge acquired through seeing may seem difficult, particularly for non-ape species, from this series of studies by Povinelli and colleagues. However, in their studies, the participants had to infer what the knower actually saw behind the occluder. A more direct demonstration of "seeing" behavior may lead to a different conclusion.

We tested tufted capuchin monkeys in a slightly modified procedure (Kuroshima et al. 2002, 2003). The hider hid a piece of food in one of three opaque containers behind a blind screen. The other two actors stood back-turned. The two actors came up to the monkeys in sequence. The knower lifted his/her own side of each container and looked inside in a random order. The guesser simply touched each container in exactly the same order as the knower. Finally, the two actors touched the containers they "believe" to be baited and the monkeys were allowed to reach toward one of the containers. The actors immediately opened the reached container and gave the food inside only if he/she found it. The order of the two actors' performances was fixed at first but was random in later sessions. All four monkeys learned to reach toward the container that the knower touched (Kuroshima et al. 2002).

In the next series of experiment conducted after a break, two monkeys were trained for generalization to five new containers having different shapes and colors. Only one of the monkeys, named Kiki, was successful in this training. This monkey was further tested with five new containers. This second set of containers not only had different shapes and colors but required different methods to open (Fig. 5). The first one was a rectangular box with a hinged top. The second was opened by lifting the top off. The third (a pot) was opened by twisting the top off. The fourth was a drawer chest. The last was a cylinder with no top, thus the inside could be checked by deliberately looking into the tube. Figure 5 shows the result of five test sessions for each container. As shown, this monkey nicely generalized her performance to all the new containers and corresponding human acts.

A problem so far was that the actions of the guesser remained the same, touching each container, and it was possible that the monkey simply learned to ignore

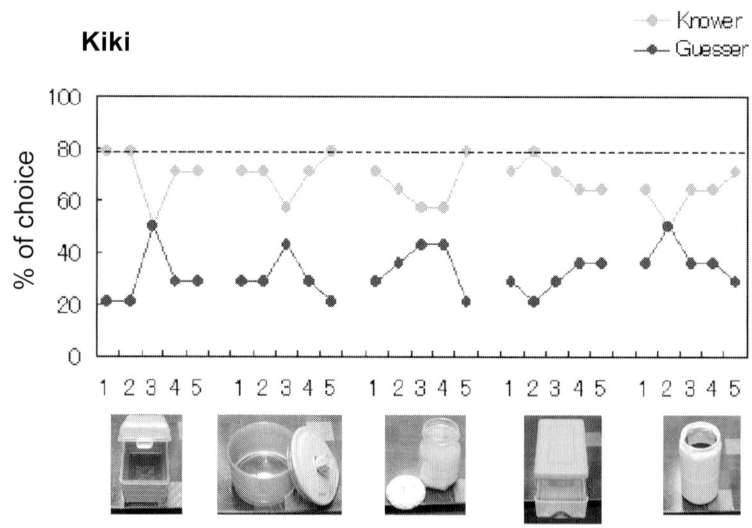

FIG. 5. Results of the generalization tests of a capuchin's discrimination of the knower and the guesser to five novel containers that provided different ways of checking the contents of the containers. The vertical axis shows the proportion of the monkey's reaching toward the knower's and the guesser's "advice." (From Kuroshima et al. 2003)

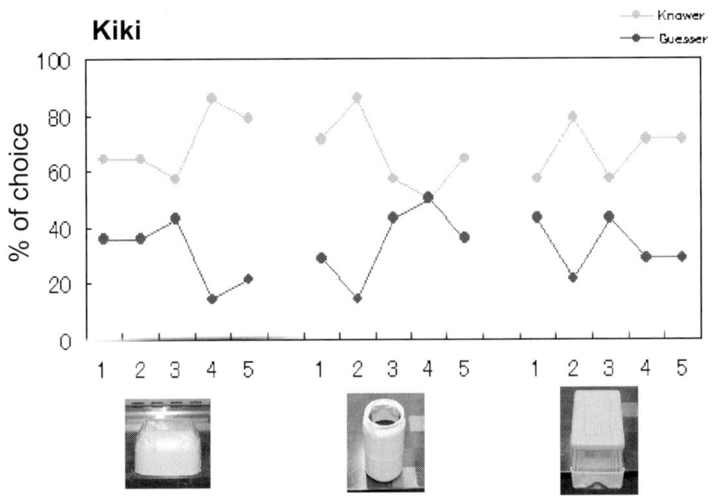

FIG. 6. Results of the generalization tests of a capuchin's discrimination of the knower and the guesser to the situation in which the guesser's behavior approximated the knower's. Other items as in Fig. 5. (From Kuroshima et al. 2003)

this actor's "advice." In the final test of this monkey, therefore, we varied the actions of the guesser. Two containers, the drawer chest and the cylinder, were used. For the drawer chest, the knower opened one of the drawers and looked into it as before. The guesser also opened the drawer but looked into another closed drawer chest. For the cylinder, the knower looked into each tube but the guesser simply made three bows. Thus, the magnitude of the action approximated between the knower and the guesser but the focus of attention differed between the two actors (Fig. 6). As a result, this successful monkey followed the knower's advice significantly more often than by chance (Kuroshima et al. 2003).

These results demonstrate that at least some capuchin monkeys are capable of recognizing knowledge states of humans suggested by their actions. This species thus understand that seeing leads to knowing.

4.4. *Learning from Other's Failure*

Learning socially from other's behavior is useful to adapt to both physical and social environments shared by the observer and the demonstrator. Previous studies on this aspect focused on imitative processes, by which an observer learns the same behavior as the demonstrator.

Copying others' actions is of course important in many situations but may not be always adaptive. For example, we should not blindly imitate a crazy driver who ignores the traffic signal to cause a severe accident. Thus, in some cases NOT copying a demonstrator can be more adaptive. We tested whether tufted capuchin monkeys could learn correct actions by observing a conspecific's failure (Kuroshima et al., in press).

We prepared two transparent boxes that looked exactly the same. One of the boxes opened from the top lid and the other from the bottom (Fig. 7). We trained three monkeys, two adults and one juvenile, to open both boxes by using the two ways above. As the boxes are indistinguishable, the monkeys sometimes succeeded by the first attempt but at other times they had to correct their responses after the first attempt.

They were then tested in pairs facing to each other. The experimenter first placed one of the boxes in front of one monkey. If the monkey opened the box on the first attempt, the trial ended. If the monkey did not open the box on the first attempt, the experimenter moved the box to the other monkey. If this second monkey opened the box on the first attempt, the trial ended. If the monkey did not, the experimenter retrieved the box and the trial ended.

Successful performance was observed only when the two adult monkeys were paired; that is, accuracy of the first attempt of the second monkey was higher than

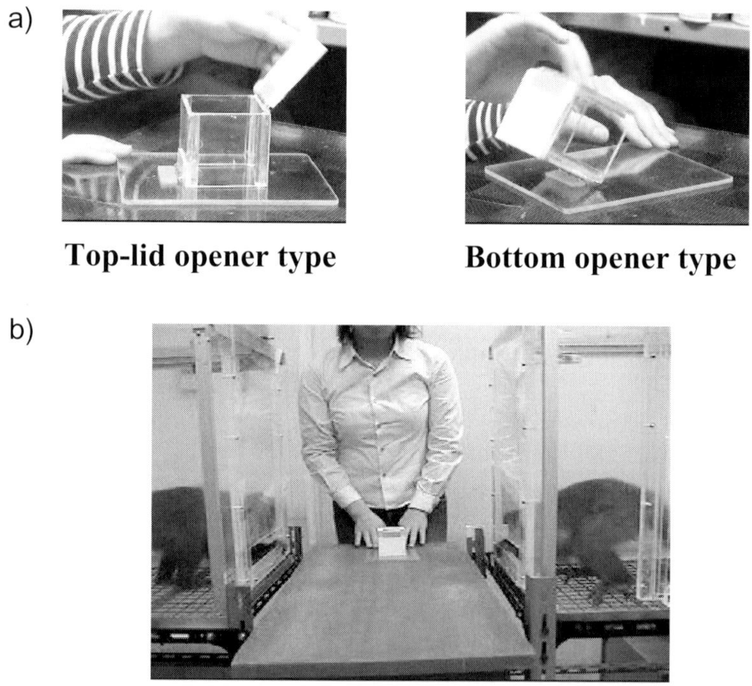

Fig. 7. **a** The two boxes used in the experiment that tested whether capuchins correct their behavior based on the other's failure. The two boxes looked identical, but the one on the *left* opened from the top lid and the one on the *right* opened from the bottom. **b** The experimental setup. Two monkeys faced each other and the experimenter moved the box toward one of the monkeys first. See text. (From Kuroshima et al., in press)

50% in this pair. Figure 8 shows how they tried to open the box in their first attempt. The third bar for each monkey shows the proportion of opening from the top and opening from the bottom in their first attempt as the first performer. The left two bars denote the same proportion in their first attempt as the second performer after the first monkey's failure.

Pigmon reliably switched his opening method depending upon Kiki's failure (left two bars). In particular, he tended to open the box from the lid, which was not his dominant method in his behavior as the first performer (third bar), after Kiki failed by opening from the bottom. Kiki showed the same tendency but her switching did not reach a statistical significance due to the small N of Pigmon's failure from the lid (left bar). However, Kiki also reliably tended to open from the lid after Pigmon failed from the bottom (second bar). Thus, the overall tendency of this pair was to correct their behavior based on the partner's failure.

It may be also interesting that these two adults did not show the same correction tendency after they observed failure by the juvenile, which might suggest that the monkeys did not trust the behavior of this young boy as a reliable sign showing an incorrect way of opening.

The results, that the two adults succeeded to take the correct action after the partner's failure, suggest that capuchin monkeys are capable of modifying their behavior dependent upon the behavior of others. In doing this, they might further take others' potential into account.

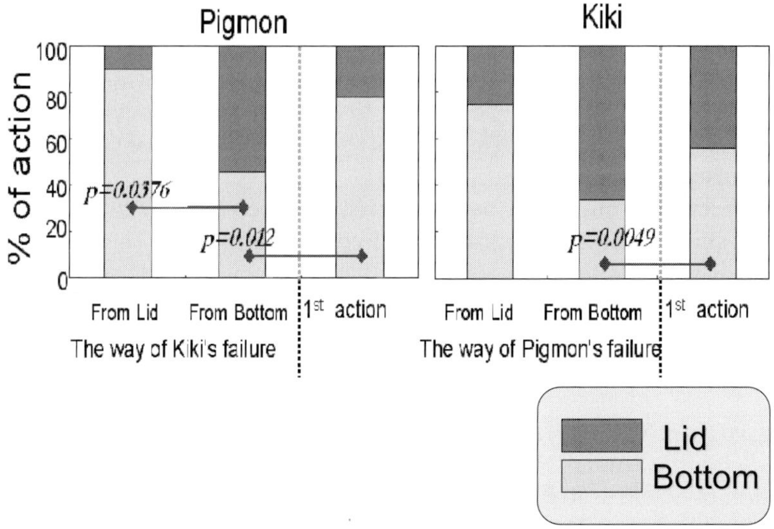

FIG. 8. Proportion of actions to open the box for each individual of the pair. The *left two bars* on each graph show the actions as the second performer and the third bar denotes the action as the first performer. Because the two boxes looked identical, the proportion of opening ways in the third bar suggests a simple behavioral preference. (From Kuroshima et al., in press)

5. Conclusion

We have described six series of experiments conducted to examine how socially intelligent tufted capuchin monkeys are. In the first series, we demonstrated that capuchins may take acts interpretable as tactically deceiving a dominant conspecific opponent who competes with themselves for food. In the second series, we have shown that individuals of this species may cooperate with each other by dividing a sequence of acts between the participants. They continue cooperating even if only one participant is rewarded at a time if they are allowed to switch the two roles.

In the remaining four series of experiment, we analyzed more fundamental cognitive abilities to complete the two most complex social interactions by reading the other's mind, or mentalizing. In the first series, we have demonstrated that capuchin monkeys are sensitive to the subtle states of human eyes. However, this recognition does not become apparent as differentiated pointing. In the second series, we have shown that this species infers the presence or absence of food based on the nature of the food source and the location of another monkey's previous visit.

In the third series, capuchin monkeys were demonstrated to recognize knowledge states of human trainers. The best performer generalized her discrimination of the knower and the guesser to a variety of new containers, new behaviors of the knower, and finally to those whose behavioral difference was only in the direction of looking, or attention. In the last experiment, adult capuchins tended to mutually correct their own behavior based on the other's failure.

All these results suggest that the deceptive and cooperative behaviors in the first two series of experiment are more likely to be based on monkeys' mentalizing rather than a consequence of simple operant learning. At the least, they have several underlying cognitive abilities to intently deceive or cooperate, sometimes altruistically, with conspecific others.

This observation might not be so surprising if we consider their virtuosity of tool-using skills as the social intelligence hypothesis suggests. The present studies suggest that a certain level of social intelligence incorporating mentalizing may be traced back to the common ancestor between humans and New World monkeys who lived more than 30 million years ago.

Acknowledgments. Preparation of this chapter was supported by the Grants-in-Aid for Scientific Studies Nos. 13410026 and 17300085 from Japan Society for the Promotion of Sciences to K. Fujita and by the 21st-Century COE Program, D-10, to Kyoto University. The capuchin monkeys were provided by the Primate Research Institute, Kyoto University, by way of its Cooperative Research Program.

References

Anderson JR, Sallaberry P, Barbier H (1995) Use of experimenter-given cues during object-choice tasks by capuchin monkeys. Anim Behav 49:201–208

Baillargeon R (1993) The object concept revisited: new directions in the investigation of infants' physical knowledge. In: Granrud CE (ed) Visual perception and cognition in infancy. Erlbaum, Hillsdale, NJ, pp 265–315

Byrne R (1995) The thinking ape: evolutionary origins of intelligence. Oxford University Press, Oxford

Byrne RW, Whiten A (eds) (1988) Machiavellian intelligence: social expertise and the evolution of intellect in monkeys, apes, and humans. Clarendon Press, Oxford

Coussi-Korbel S (1994) Learning to outwit a competitor in mangabeys (*Cercocebus torquatus torquatus*). J Comp Psychol 108:164–171

Epstein R, Lanza RP, Skinner BF (1980) Symbolic communication between two pigeons (*Columba livia domestica*). Science 207:543–545

Fragaszy D, Izar P, Visalberghi E, Ottoni E, de Oliveira MG (2004) Wild capuchin monkeys (*Cebus libidinosus*) use anvils and stone pounding tools. Am J Primatol 64:359–366

Frith U, Morton J, Leslie AM (1991) The cognitive basis of a biological disorder: autism. Trends Neurosci 14:433–438

Fujita K, Kuroshima H, Asai S (2003) How do tufted capuchin monkeys (*Cebus apella*) understand causality involved in tool use? J Exp Psychol Anim Behav Process 29: 233–242

Fujita K, Kuroshima H, Masuda T (2002) Do tufted capuchin monkeys (*Cebus apella*) spontaneously deceive opponents? A preliminary analysis of an experimental food-competition contest between monkeys. Anim Cognit 5:19–25

Hare B, Brown M, Williamson C, Tomasello M (2002) The domestication of social cognition in dogs. Science 298:1634–1636

Hattori Y, Kuroshima H, Fujita K (2005) Cooperative problem solving by tufted capuchin monkeys (*Cebus apella*): spontaneous division of labor, communication, and reciprocal altruism. J Comp Psychol 119:335–342

Hattori Y, Kuroshima H, Fujita K (2007) I know you are not looking at me: capuchin monkeys' (*Cebus apella*) sensitivity to human attentional states. Anim Cognit 10: 141–148

Itakura S (1997) Understanding of the human mind by nonhuman primates: The theory of mind in nonhuman primates. Jpn Psychol Rev 40:8–21 (in Japanese with English summary)

Itakura S, Anderson JR (1996) Learning to use experimenter-given cues during an object-choice task by a capuchin monkey. Curr Psychol Cognit 15:103–112

Jolly A (1988) Lemur social behaviour and primate intelligence. In: Byrne, RW, Whiten A (eds) Machiavellian intelligence: social expertise and the evolution of intellect in monkeys, apes, and humans. Clarendon Press, Oxford, pp 27–33

Kaminski J, Call J, Tomasello M (2004) Body orientation and face orientation: two factors controlling apes' begging behavior from humans. Anim Cognit 7:216–223

Kuroshima H, Fujita K, Masuda T (2002) Understanding of the relationship between seeing and knowing by capuchin monkeys (*Cebus apella*). Anim Cognit 5:41–48

Kuroshima H, Fujita K, Adachi I, Iwata K, Fuyuki A (2003) A capuchin monkey (*Cebus apella*) understands when people do and do not know the location of food. Anim Cognit 6:283–291

Kuroshima H, Kuwahata H, Fujita K (2004) Learning from other's mistakes in capuchin monkeys (*Cebus apella*). Anim Cognit in press.
Mendres KA, de Waal FBM (2000) Capuchins do cooperate: the advantage of an intuitive task. Anim Behav 60:523–529
Menzel EW (1974) A group of chimpanzees in a 1-acre field: leadership and communication. In: Schrier AM, Stollnitz F (eds) Behavior of nonhuman primates. Academic Press, New York, pp 83–153
Mitchell RW, Anderson JR (1997) Pointing, withholding information, and deception in capuchin monkeys (*Cebus apella*). J Comp Psychol 111:351–361
Myowa-Yamakoshi M, Tomonaga M, Tanaka M, Matsuzawa T (2003) Preference for human direct gaze in infant chimpanzees (*Pan troglodytes*). Cognition 89:B53–B64
Onishi KH, Baillargeon R (2005) Do 15-month-old infants understand false beliefs? Science 308:255–258
Povinelli DJ, deBlois S (1992) Young children's (*Homo sapiens*) understanding of knowledge formation in themselves and others. J Comp Psychol 106:228–238
Povinelli DJ, Eddy TJ (1996a) Factors influencing young chimpanzees' (*Pan troglodytes*) recognition of attention. J Comp Psychol 110:336–345
Povinelli DJ, Eddy TJ (1996b) What young chimpanzees know about seeing. Monographs of the Society for Research in Child Development, Serial No. 247, Vol. 61, No.3
Povinelli DJ, Nelson KE, Boysen ST (1990) Inferences about guessing and knowing by chimpanzees (*Pan troglodytes*). J Comp Psychol 104:203–210
Povinelli DJ, Parks KA, Novak MA (1991) Do rhesus monkeys (*Macaca mulatta*) attribute knowledge and ignorance to others? J Comp Psychol 105:318–325
Savage-Rumbaugh ES, Rumbaugh DM, Boysen S (1978) Symbolic communication between two chimpanzees (*Pan troglodytes*). Science 201:641–644
Vick S-J, Anderson JR (2000) Learning and limits of use of eye gaze by capuchin monkeys (*Cebus apella*) in an object-choice task. J Comp Psychol 114:200–207
Visalberghi E (1990) Tool use in *Cebus*. Folia Primatol 54:146–154
Visalberghi E, Quarantotti BP, Tranchida F (2000) Solving a cooperation task without taking into account the partner's behavior: the case of capuchin monkeys (*Cebus apella*). J Comp Psychol 14:297–301

2
Communication Between Mother and Infant Chimpanzees and Its Role in the Evolution of Social Intelligence

Satoshi Hirata

1. Chimpanzee Social Intelligence

Many primate species, including humans, live in societies in which they interact and communicate with other individuals during daily life. Aspects of intelligence that appear in such social lives include social intelligence (Humphrey 1976) or Machiavellian intelligence (Byrne and Whiten 1988; Whiten and Byrne 1997).

Based on chimpanzee research, Premack and Woodruff (1978) presented a groundbreaking framework for the social aspects of intelligence. They suggested the possibility that chimpanzees understand the intentions, desires, and emotions of others and predict the future behavior of others based on an understanding of their mental states. Premack and Woodruff (1978) coined the phrase "theory of mind" to characterize the cognitive ability to understand the mental state of others. The study of "theory of mind" has flourished in developmental psychological research on humans and has subsequently returned to primatology.

Tomasello and his colleagues (Call 2001; Call and Carpenter 2001; Hare 2001; Hare et al. 2000, 2001; Tomasello et al. 2003a,b) conducted a series of research to examine chimpanzees' understanding of others' visual perceptions, an elemental mental state attribution. In their experiments, a dominant and a subordinate chimpanzee were brought into two rooms on opposite sides of a third room. There were various conditions, but basically, a subordinate chimpanzee could see a piece of hidden food that the dominant chimpanzee could not see. The subordinate chimpanzees took advantage of being able to see the food that the dominant opponents could not see, indicating that the subordinates knew what the dominant chimpanzees could and could not see.

Similar procedures were used by Hirata and Matsuzawa (2001; see also Hirata 2006a) to examine social intelligence in chimpanzees. While a pair of female

---

Great Ape Research Institute (GARI), Hayashibara Biochemical Laboratories, Inc., 952-2 Nu, Tamano, Okayama 706-0316, Japan

chimpanzees was kept inside, a human experimenter entered an outdoor enclosure and hid a banana in one of five prearranged places. One of the two chimpanzees (the witness) could see where the experimenter hid the banana, whereas the other (witness-of-witness) could not directly see the hiding spot. The result was simple in the beginning; the witness went to the hiding place and obtained the reward. After repeated trials, the witness-of-witness began trying to steal the reward by following and chasing the witness. The witness opposed this action by going in a direction different from the hiding place. After the witness-of-witness followed the witness and searched this empty area, the witness returned to the correct place and obtained the reward, exhibiting deception. The witness-of-witness then acted against this deception by remaining very close to the witness. Thus, tactics and countertactics were developed during the course of the experiment, proving higher social intelligence in chimpanzees.

More naturalistic observations by de Waal (1982) also showed complex social maneuvering in adult male chimpanzees. His observations illustrated how male social ranks are established by the "political" tactics of factional disputes for higher ranks among powerful males. Groups had coalitions, alliances, interventions, and betrayals. Such struggles for power have also been observed in the wild (Nishida and Hosaka 1996).

The two kinds of experimental studies outlined above used competition over food in their experimental paradigms. Other naturalistic observations focused on chimpanzee behavior in competing for rank. As noted by Hare (2001), "Primate social life is highly competitive. This means that all aspects of primate themselves, including their cognitive abilities, have likely been shaped by the need to out-compete conspecifics." There is no doubt that winning over competitors is highly beneficial for survival and that the sophisticated cognitive abilities of primates can be seen in competitive situations. However, when we look at ourselves, humans are not always thinking about winning in competition. Sometimes we help and cooperate with each other and coordinate behavior with others. Cooperative aspects are important in human social intelligence. This raises questions as to whether chimpanzees are socially intelligent in situations other than competition. Do chimpanzees cooperate or coordinate behavior with others?

Bearing such questions in mind, this chapter considers the communication in mother–infant dyads and its role in the evolution of social intelligence by observing cases involving chimpanzees. All primate species experience their first social interactions with their parents. In the case of chimpanzees, paternity is usually unknown; thus, the parent is the mother. Infants are cared for by their mothers and cling to them until they are weaned (Matsuzawa 2006a). The interactions and communications between mothers and offspring are the first sociality that human and nonhuman primates experience after birth.

In the following sections, I describe two types of observations of captive chimpanzees. The first observations focus on mothers' transportation of their infants. The second involves a series of episodes related to a primiparous chimpanzee's delivery and rearing behavior. Through observations of these cases, I consider the social cognition underlying chimpanzee behaviors and the evolutionary background of social intelligence or the ability to communicate with others. Various researchers have described chimpanzee deliveries and subsequent mother–infant interactions (Bard 1994; Bloomsmith et al. 2003; Davis et al. 1981; Horvat and Kraemer 1981; Horvat et al. 1980; Nankivell et al. 1988; Nicolson 1977; Rogers and Davenport 1970; Struthers et al. 1990), but there is still much to learn from these events.

2. Observations of Mothers in the Wild

Chimpanzee and other great ape infants, including human infants, develop more slowly than other non-ape species (Purvis et al. 2003). Weaning in chimpanzees takes 3 to 5 years; until that time, infants are often carried by their mothers (Hiraiwa-Hasegawa 1990; van de Rijt-Plooij and Plooij 1987). Infants younger than 3 months of age are barely able to move by themselves and mostly cling to their mothers. Chimpanzees spend about half of their day in trees and travel from one tree to another searching for food including fruits, leaves, and piths (Takemoto 2004). Mothers travel arboreally or terrestrially, and infants cling tightly to their mothers as they are transported. After 3 months of age, infants gradually become able to move by themselves, but it takes years until they can travel completely alone. Until 3 to 5 years of age, young chimpanzees may sometimes travel alone, but at other times they cling to their mothers for assistance.

When I observed a group of wild chimpanzees, I sometimes witnessed interactions in which a mother would hold out her hand to her offspring. For example, if a mother had easily moved from one tree branch to another but the gap between the branches was too wide for the infant, the mother would look back at the infant and stretch out her hand. The infant would take the mother's hand, and the mother would pull the infant to the next branch. This sequence of behavior can be considered a reflection of chimpanzees' social intelligence in the sense that the mother understood that the gap was too wide for the infant to cross. The mother's stretching out of her hand constituted a communicative sign to the infant. The infant responded to the mother's sign and was helped by the mother to move to the next branch. This is a very simple behavioral sequence but shows a kind of communication in which one individual behaves toward another and the other individual responds to it.

I also observed a group of 150 free-ranging, provisioned Japanese macaques for 8 months in Kyoto, Japan. The group included several pairs of mothers and newborn infants. To my recollection, I never observed cases in which a mother macaque extended her hand to help her offspring. Rather, infants caught the mother when the mothers started to move, or the mother pulled the infant before moving. Thus, the interaction was one directional. Chimpanzee mothers' way of helping their offspring was more conspicuous after my observations of the Japanese macaques.

3. Observation of Three Mother–Infant Pairs in Captivity

3.1. Parturition and Maternal Behaviors

In 2000, three chimpanzees gave birth to offspring at the Primate Research Institute of Kyoto University, Inuyama, Japan (Matsuzawa 2006b; Matsuzawa and Nakamura 2004). A 23-year-old chimpanzee named Ai gave birth to a male named Ayumu in April, a 19-year-old named Chloe gave birth to a female named Cleo in June, and a 16-year-old named Pan gave birth to a female, Pal, in August.

Many captive chimpanzee mothers fail to show appropriate maternal behaviors and almost 50% reject care of their babies (Brent et al. 1996), indicating that maternal behavior is not completely genetically programmed in chimpanzees. Maternal behavior is greatly influenced by environment, experience, and learning before the chimpanzee becomes a mother (Bloomsmith et al. 2003).

Ai came to the Primate Research Institute when she was about a year old and was then cared for by humans (Matsuzawa 2003, 2006b). Chloe was born at a zoo but her mother did not take care of her, and she was reared by humans. Pan was born at the Primate Research Institute, but her mother also did not take care of her, and she was reared by humans. Considering the histories of these three mothers, there was concern that they might not accept and care for their babies. Personnel at the Primate Research Institute tried to improve the possibility that they would care for their babies by showing them videotapes in which wild mothers cared for infants or by urging them to hold a chimpanzee infant doll for practice (Matsuzawa and Nakamura 2004).

It is difficult to determine whether or to what extent such efforts had an effect. There were individual differences in their behaviors, but none of the three mothers showed ideal maternal behavior. Ai embraced her infant immediately after delivery, but frequently held the infant in inappropriate manners. On the first night after delivery, she held the infant upside down or between her thigh and abdomen. She continued to hold the infant in unstable positions in the early days after giving birth. Ai's inappropriate behaviors gradually ceased, and after approximately a week she began to hold the infant in a stable position for 24 h.

Chloe, however, moved away from her newborn baby immediately after delivery. She was then carefully observed by human researchers and caretakers for a while, but she showed no signs of holding her baby spontaneously. Pan lay down beside her baby but showed no signs of holding the infant. Personnel familiar to these mothers stayed with each of them to prompt them to hold their infants. Eventually, the mothers held their babies, but Chloe had another problem. She did not suckle her infant. She rejected the infant's attempts to suckle by tearing the infant away when she tried to contact her nipple. Again, people familiar to Chloe attempted to prompt suckling, and gradually Chloe allowed the infant to nurse.

The three mothers' maternal behaviors were unstable, as just described, but human efforts probably improved the situation, and the mothers acquired necessary maternal behaviors. Besides the human efforts, improvements also reflected increased communication between the mothers and infants in which the infants vocalized and the mothers responded to the vocalizations. For example, when an infant whimpered a "hoo" sound when she could not reach the nipple, the mother responded and repositioned the infant. Infant whimpering when separated from the mother also caused the mother to respond and rush to hold the baby. Communication evoked by infants' whimpering or screaming was often seen in the early stages of the mothers' baby care.

3.2. Communication Before Traveling

The infants gradually moved farther from their mothers as their motor abilities developed, enlarging the area in which they could move by themselves. Ayumu was first observed walking quadrupedally by himself apart from his mother when he was 18 weeks old. The same behavior was observed when Cleo was 23 weeks old and Pal was 25 weeks old. It was not easy for them, however, to move freely around the enclosures where they were housed. The enclosures of the Primate Research Institute have a 15-m-high metallic climbing structure, ropes, platforms at various heights, and planted trees, functionally simulating an African forest (Ochiai and Matsuzawa 1997). The mothers traveled freely in the enclosures here and there, up and down, but the small infants could not do the same.

In such situations, the three mothers were observed communicating with their infants in the same ways that wild mothers helped their infants with traveling (Fig. 1). For example, a mother stretched out her hand toward her offspring who was away from her before traveling. The infant then approached the mother to take her hand. The mother cradled the infant and moved from one place to another while carrying her offspring. In another example, a mother stooped down showing her backside to the offspring, and the infant then rode on the mother's back as the mother traveled. In a third example, a mother poked the body of her infant who was a few inches away from her. The infant

a b

Fig. 1. A chimpanzee mother stretching out her hand to her offspring (**a**) and touching the body of the infant (**b**) before starting to travel a distance.

responded to this prodding and went to ride on her mother. Mothers' behaviors such as outstretching their hands, stooping down to show their backsides, and poking the body of the infant can be recognized as communicative signals toward the infants. The mothers emitted signals to the infants to which the infants responded, allowing the mother to carry the infant so that they could travel together.

I analyzed the behaviors of the three mothers and their infants when they started to travel from one place to another in an enclosure of the Primate Research Institute from May to August 2001 (Ayumu was 12–15 months old, Cleo was 10–13 months old, and Pal was 9–12 months old). I considered cases in which the following conditions were met. First, a mother and infant stayed apart when the mother started a travel. The mother and infant then made physical contact in some manner. Finally, the mother traveled while carrying the infant. The enclosure in which the video recording was conducted was about 150 m^2 in area and adjacent to the larger 700-m^2 enclosure. The smaller enclosure had a 6-m-high climbing structure, ropes, trees, and a stream. A digital video camera was used for the recordings, which took place at around noon for 1 h once every week for each of the three infants.

Behavioral events that occurred when physical contacts were made between mothers and infants who were initially apart could be categorized into three cases. In the first case, called unilateral infant-to-mother, the infant caught the mother. In this case, a mother had started traveling alone and had not shown any behaviors toward her infant. The infant moved toward the moving mother and caught her. For the second case, termed unilateral mother-to-infant, the mother pulled her infant. In this case, an infant showed no behavior toward the mother until the mother pulled the infant. In the third case, called reciprocal, both the mother and her offspring showed some kind of behavior toward each other before their physical contact. These cases are exemplified by a mother stretching out her hand to her infant and the infant approaching the mother to grab the mother's hand.

Table 1 shows the proportion of these three categories observed in the three mother–infant pairs, Ai and Aymu, Chloe and Cleo, and Pan and Pal. In all three pairs, one-third to one-half of the interactions were reciprocal, in which the three mothers and their infants communicated in some way to make physical contact before they traveled together.

TABLE 1. Percentages of the three types of behavioral patterns in the three mother–infant pairs.

	Type of behavioral patterns		
Mother–infant pair	Unilateral, infant-to-mother (%)	Unilateral, mother-to-infant (%)	Reciprocal (%)
Ai–Ayumu	24	14	62
Chloe–Cleo	23	43	34
Pan–Pal	11	30	59

3.3. Maternal Judgments

The mothers did not necessarily carry their infants when they traveled across the enclosure. They sometimes left the infant and traveled alone. The mothers seemed to judge the situation to decide whether they should carry the infant or travel alone. The infants were more or less able to move by themselves, but their motor skills were still immature. If a mother moved far from her offspring, the infant could not catch up with her. When the infant and mother were separated by such distances, the infant emitted a vocalization called a whimper. The mother responded to the whimper and returned to the infant. There were several different cases in which infants had difficulties catching up to their mothers. In some cases, infants could not manage ascents and descents in the travel route, overly wide gaps, or distances that were simply too long. In the following analysis, the relationship between the travel distance and the mothers' behaviors were examined to consider the mothers' judgments depending on different situations.

The same data described above were again analyzed to determine the frequency of mothers carrying their infants as a function of subsequent travel distances (Fig. 2). Travel distances were divided into three categories: 0–1, 1–5, and more than 5 m. The proportion of infant carrying was calculated by dividing the number of times the mothers carried the infant for each travel-distance category by the number of times the mothers traveled the same travel distance with or without carrying the infant. The results indicated that the mothers carried their infants more frequently as the travel distance became longer (Page test, P = 0.014). The same tendency was found when the relationship between the proportion of infant carrying was calculated as a function of the vertical elements of the distance traveled (Page test, P = 0.005). The mothers carried the infant more often if there were more ascents and descents along the travel route.

The foregoing cases were further analyzed to examine physical contact initiatives by the mothers related to carrying the infants. Figure 3 shows the proportion of times each mother contacted her infant before starting a travel as a function

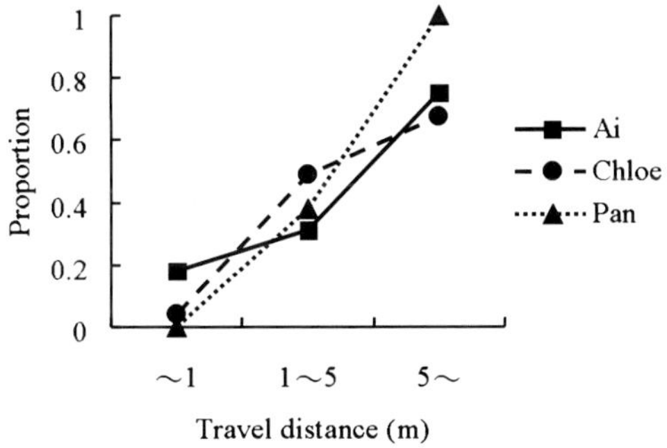

FIG. 2. The proportion of the mothers carrying their infants as a function of travel distance.

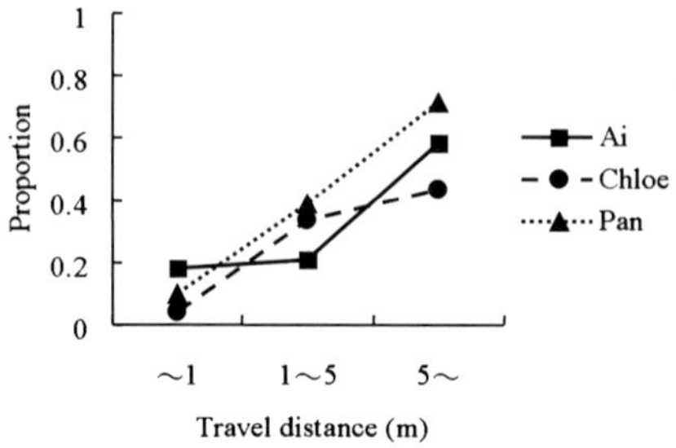

FIG. 3. The proportion of times mothers worked with their infants so as to carry them before starting to travel.

of subsequent travel distance. Cases in which mothers initiated "reciprocal" and "unilateral mother-to-infant" interactions were combined to represent cases in which mothers worked with their infants before beginning travel. The results indicated that when mothers' travel distances were less than 1 m, the mothers rarely showed any behaviors toward the infant and traveled alone. However, when the mothers traveled longer distances, they initiated and accomplished physical contacts with the infant more often (Page test, $P = 0.014$); these contacts included pulling the infant or extending an arm to the infant before starting to move. The mothers then traveled with the infant by carrying them.

These results suggest that the mothers had perhaps already decided on the goal of the travel before they started moving. When the goal was nearby, they left the infant and moved alone. When the goal was far, the mothers initiated contact with the babies before starting travel, prompting the baby to cling to the mother; the mother and infant then traveled together, with the infant being carried. The mothers seemed to take their infants into consideration and decided whether they should carry them depending on the distance to their planned travel goal.

In some other cases, the infant showed the first behavior toward the mother and the mother responded to it. For example, an infant moved away from the mother via ropes or by crossing a small gap but could not go return by itself. The infant then extended its hand to the mother, and the mother responded by pulling the infant's hand. Other examples include the following. At the Primate Research Institute, chimpanzees are called to the experimental rooms from an enclosure through tunnel passageways. There are several electrically controlled gates along the passageway. The gates are opened when a chimpanzee passes through and closed after the chimpanzee passes. Infant chimpanzees occasionally extended hands and feet, not caring about the closing gates. When the hands or feet of infants were about to be caught in the gates, the mothers noticed and pulled the infants to a safer place. The mothers monitored the infants carefully and understood the situation the infants faced. The mothers made appropriate judgments and assisted the infants when necessary.

It is unlikely that these maternal behaviors are genetically programmed, as shown by the fact that half of the captive chimpanzee mothers did not take care of their offspring in general and the three study chimpanzees, Ai, Chloe, and Pan, lacked appropriate behaviors when caring for their newborn babies. There were various situations in which the mothers had to assist their infants in their daily lives. These situations changed according to infant development. It is almost impossible that each behavior in each situation was genetically programmed. What leads to the emergence of appropriate chimpanzee maternal behaviors? I believe it is social intelligence. The chimpanzee mothers communicated with infants through several types of behaviors, comprehended the state of the infants, and planned future behaviors taking other individuals into consideration. Infants, on the other hand, understood the behavioral signs of their mothers and responded appropriately. These social abilities in chimpanzees typically appear in situations such as a mother helping an infant to travel (see also Bard, 1992 regarding cases for orangutans).

4. A Case Report of a Young Primiparous Chimpanzee

4.1. Parturition

I had another opportunity to observe a young chimpanzee female giving birth. A description of this event helps clarify the process of a chimpanzee becoming a mother. On May 31, 2004, Tsubaki had her first menstruation. Tsubaki was a female chimpanzee housed at the Hayashibara Great Ape Research Institute, Japan. She was born on February 17, 1996 and was thus 8 years and 3 months old at the time. This age may be slightly younger than average for the first menstruation in chimpanzees, but is probably within the range of normal sexual maturation. She then menstruated almost once every month at regular intervals. Ovulations were also confirmed from urine samples, about 2 weeks before each menstruation. She showed tumescence of the sexual skin during this period, and two males of the same group repeatedly mated with her, accompanied by ejaculation. In November of the same year, 6 months since her first menstruation, a scheduled menstruation was not observed. A pregnancy test of her urine sample revealed that she had become pregnant. The fetus was confirmed using four-dimensional (4-D) ultrasonography (Takeshita et al. 2006).

On July 8, 2005, at 0300 on the 248th day of gestation, Tsubaki showed signs of being in labor. At midnight of that day she had awoken, moved, and sometimes assumed a posture of bearing down. To prepare for delivery, she was separated from other group members and brought to a single room. She began to bear down at intervals of 30 min to several minutes. At 0922, she gave birth to a female chimpanzee named Natsuki.

In advance of the delivery, Tsubaki showed an unusual behavior. She appeared to ask a human staff member at the institute for assistance in delivery. Staff of the institute stayed in the same room with Tsubaki to monitor the delivery. One of the staff members had been with Tsubaki since she was 3 years old. While bearing down, Tsubaki suddenly took the foot of the human staff member and pressed it to her abdomen. She seemed to be trying to push out the baby using the staff member's foot.

Tsubaki had been reared by her mother until she was 3, but she did not see other chimpanzee infants. As expected by such a rearing history, after giving birth, Tsubaki did not hold the infant in an appropriate position, although she did hold the infant. The mother frequently put the newborn infant into a position between her thigh and abdomen or under her underarms. In such positions, the helpless infant, Natsuki, could not reach her mother's nipples. Because Natsuki could not drink her mother's milk for a day and a half after birth, the human staff decided to assist the mother and infant. Staff members tried to move the infant gently to a breast-feeding position, or held Tsubaki's hands and feet so that the infant could easily move to the breast. Additionally, the baby was given 5% glucose liquid to prevent dehydration. The infant had more chances to nurse during this time. However, the mother seemed to dislike the infant's suckling, probably because she was unfamiliar with the feeling. Thus, when Natsuki approached her nipple, Tsubaki pushed the infant away.

Human staff members again tried to reduce Tsubaki's rejection of suckling. When Tsubaki tried to push her infant away from her nipple, staff members took the mother's hand and prompted her to be patient. By having several staff members take such measures for entire days, the infant was first able to nurse by herself on the third day after birth. However, the suckling behavior was erratic, and Tsubaki still tended to reject nipple contact. During the daytime when Tsubaki was awake, she prominently rejected sucking by the infant. However, at night when she fell asleep, Tsubaki continued to sleep or was less concerned even if she woke up when Natsuki searched for a nipple. Nighttime was a good time for the infant to nurse, and the consumption of mother's milk mostly occurred at night in the beginning. In time, Tsubaki's tendency to reject suckling gradually decreased, and after about a week from giving birth, the infant could stably drink milk from her mother (Fig. 4). Soon after that, Tsubaki

FIG. 4. Tsubaki and Natsuki.

actively suckled the infant by repositioning the infant closer to a nipple when the infant began to move in search of nursing. The mother seemed to understand what was needed for the infant.

4.2. Removing the Infant

Wild chimpanzee mothers hold their newborn infants for 24 h a day. The infants begin to move away from mothers at about 3 months of age. Until that time, the infants always cling to their mothers. Newborn babies are not strong enough to cling to their mothers by themselves; thus, the mothers cradle their infants to support them (Matsuzawa 2006a). This pattern was different for Tsubaki. About 2 weeks after giving birth, Tsubaki began to show signs of placing Natsuki away from her body. The infant grabbed her mother's fur and skin with both hands, but Tsubaki took the infant's hands and tried to remove them and place the infant on the floor. At that moment, the infant cried, and Tsubaki held the infant again. Thus, Tsubaki did not actually place the infant apart from her on this first attempt.

In time, however, Tsubaki showed even clearer tendencies to pull the infant away from her body. At the same time, the sleeping infant occasionally naturally fell off her mother's body. The mother's suckling behavior had become stable by this time; thus the infant may have become relaxed, which may have caused such incidents. After drinking milk and falling asleep, the infant's hands holding onto the mother became looser and began to fall from the mother, resulting in the infant's falling onto the floor. There was straw on the floor, and both the mother and infant would lie on the straw. In the third week after birth, however, Tsubaki began to pull her infant away deliberately.

4.3. Cry and Hold

Tsubaki frequently pulled her baby away from her body. She left the infant and rested some distance away. Natsuki lay down on the floor on her back if she was relaxed or sleeping. At first, she could not turn over by herself. When she awoke and became uneasy, she moved her hands and feet while lying on the floor. After some time, Natsuki would utter a "hoo" whimper call, with her voice gradually becoming louder and turning into a scream or cry. On hearing the cry, Tsubaki would rush to the infant and embrace her.

If the infant cried, the mother came to hold her baby. This was consistent and customary behavior observed in this young mother. Tsubaki's maternal behaviors were not perfect from the beginning, but she always rushed to hold the baby when she heard the infant crying. Crying was the only and most powerful communication tool for the helpless infant. It is unclear what the mother understood the crying to mean, but she always returned to cradle her infant when Natsuki cried, and the infant stopped crying immediately after she was picked up by her mother (c.f. Bard 2000).

4.4. Depositing the Infant

Tsubaki performed another strange behavior. She deposited the infant with human staff members. As described above, the staff members had stayed in the same room with Tsubaki and had monitored the mother and baby. Meals were given by one of the human staff in rotation, and after the meal the staff member continued to stay with mother and baby. When the mother finished eating her meal and the infant was satisfied after drinking milk from her mother, both became relaxed. The mother placed the infant apart on the straw floor. She picked up the baby again if the infant cried but left the infant on the floor if she did not cry. The staff member examined the infant when she was separated from her mother. Following these experiences, Tsubaki began to push the infant to the side of the human staff member, who quite naturally held the baby.

Once such an event happened, the same action was repeated almost every day. After the meal and before falling asleep, Tsubaki deposited the baby with the staff member. The baby did not care who held her and was quiet when the human cradled her. The way in which Tsubaki deposited her infant became clearer with time; she took the infant's hands and feet, pulled them from her body, and gave the infant to the staff member (Fig. 5). If the infant did not release her mother's body, Tsubaki showed a grimacing facial expression displaying distress and tried very hard to deposit the infant with the human. If the infant cried, the mother came back to receive the infant; if the mother fell asleep when the baby was apart from her, then the staff member gently gave back the infant to the mother.

Tsubaki was separated from other conspecific group members after delivery, but both the mother and infant seemed fine. They thus gradually rejoined the group. Encounters with their chimpanzee companions were repeated every day, and after some trials the infant was accepted by other group members. Tsubaki, although she deposited her baby with human staff, never handed the infant to other chimpanzees. Rather, if other chimpanzees came close, she avoided them as if to protect her baby. She appeared to consider other chimpanzees as possible

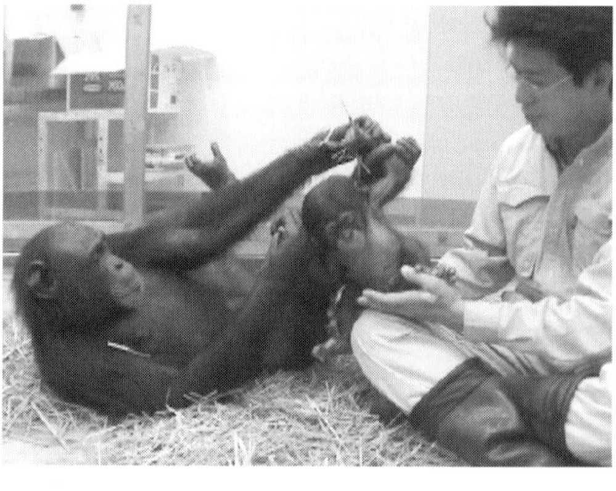

Fig. 5. Tsubaki handing her infant to a human.

threats to her baby. When there were other chimpanzee companions, Tsubaki always held her baby. It may be that she did not abandon caretaking when she handed her baby to the staff members; rather, she may have grasped the need to protect her baby and asked the human staff to help care for the baby.

Tsubaki also invented another tactic to carry the infant on her back. Wild chimpanzee mothers begin to carry their infants on their backs when the infants are about 2 months of age. Before that period, the infants cannot grab the mothers tightly and their mothers hold them ventrally and support them with their hands. Tsubaki, however, began to carry Natsuki on her back when the infant was less than 1 month old. At this age, the infant could not move or grasp the mother freely. Tsubaki seemed to adopt the back-carrying technique to avoid the infant's clinging to her abdomen. Natsuki sometimes moved around the breast to search for a nipple or grasped a bundle of fur or skin when she tried to change her position. Tsubaki probably disliked these movements. When the infant was on Tsubaki's back, she could not search for a nipple. The mother seemed to use this tactic when she walked around to keep the infant still. However, Tsubaki would still bring her infant to her abdomen when necessary and suckle her. She also never dropped the infant. She supported Natsuki with her hands when the infant became unstable. The mother seemed to make appropriate judgments depending on the situation.

4.5. Assessing the Infant

Tsubaki showed another interesting behavior toward her infant. She sometimes seemed to test the infant. For example, Tsubaki would leave the infant on the floor. The infant could not turn over by herself and thus lay on her back and moved her hands and feet. The mother would come back to the infant and turn her over. Sometimes the behavior would finish at this point but other times another behavior would follow. In this case, after turning the infant over, Tsubaki would remain some distance from the infant and clap her hands while watching the infant or emit a grunt-like vocalization while bowing to the infant. It was as if she was testing how the infant would respond. Because the infant could not move by herself, she would cry at some point, at which time Tsubaki would return and embrace the infant.

Tsubaki had never seen newborn babies. This was her first experience with a small chimpanzee who could not turn over, move, or walk. Her companions in the same group all walked quadrupedally and sometimes bipedally. She had encountered no chimpanzee who could only lie down. When Tsubaki turned her infant over and clapped her hands while watching Natsuki from a distance, or emitted a grunt-like vocalization while bowing to Natsuki, Tsubaki seemed to be checking the infant's response to see if the small chimpanzee would move and walk by some stimulation. It may be that she not only accepted the infant passively but also tried to understand the infant positively by exploring the infant's responses (see Nicholson 1977, and Bard 1994, for similar observations, and also Whiten 1999 for another case involving a gorilla).

Tsubaki's behaviors such as rejecting suckling, placing the infant apart from herself, and depositing the infant with humans are reminiscent of "parenting distress" in human mothers. In the case of humans, surrounding people support mothers to reduce the mothers' load. In other words, humans need support to deliver and take care of infants. Perhaps the same is true, at least partly, for chimpanzees, as suggested by Tsubaki actually asking for support from surrounding humans from the beginning of delivery. Wild chimpanzees have chances to learn how to handle babies during their juvenile and adolescent periods, and they do not need actual support when they become mothers themselves. Tsubaki did not have such learning opportunities and thus it is not strange that she showed some imperfect maternal behaviors. However, she seemed to explore solutions to these problems through her direct trials and errors with her infant.

5. Mother–Infant Interactions and Social Intelligence

During the course of primate evolution, infant development has become more prolonged (Parker and McKinney 1999). As I observed in the field, Japanese macaque newborns are able to walk independently, apart from their mothers, within several days after birth. In contrast, it takes about 3 months for chimpanzee newborns to move quadrupedally. Human neonates need 8 to 11 months to be able crawl by themselves. As a result of such slow infant development, the degree to which mothers must care for their babies increases.

Precocious animals such as deer and horses are able to move by themselves and eat the same foods as adults soon after birth. The mothers of these animals do not need to carry their infants after birth. Altricial chicks cannot fly as adult birds can, but they stay in nests or restricted areas so that they do not have to move. These mothers also do not need to carry their infants. In contrast, the majority of primate infants cannot move freely and are not protected in nests as neonates. The solution is that the mothers of many primate species carry their infants (Ross 2003). The infants of great apes and humans are especially helpless and develop very slowly, and the need for maternal care is great. Chimpanzees and other species travel both on the ground horizontally and also move from tree to tree in a three-dimensional space; such a wide activity space is even more difficult for infants to move across on their own. The infants thus require help from mothers, and mothers need to help their infants move. For the reproductive success of mothers and survival of infants, it would be important to have smooth communication to coordinate traveling together, depending on the situation. This need suggests the reason for the frequent communication associated with traveling between mothers and infants.

Understanding others' intentions, desires, and emotions, interpreting and predicting others' behaviors, and engaging in complex interactions are typical elements of social intelligence. The role of social intelligence has often been discussed in connection with social maneuvers among adult male chimpanzees or deceptive behaviors that bring merits to the self (Whiten and Byrne 1988).

Experimental studies focusing on competition over food have illustrated social intelligence in chimpanzees (Hare 2001; Hirata 2006a). But do these studies mean that chimpanzees show social intelligence only to win competitions over resources? Cooperative or friendly elements of social intelligence have been identified in chimpanzee mother–infant interactions, the first social interactions that every individual experiences after birth. Several features of their behaviors and interactions have been clarified by various studies. For example, neonate chimpanzees smile during REM sleep, and later begin to smile socially at their peers (Mizuno et al. 2006; Tomonaga et al. 2004). Mutual eye gazing also occurs between mothers and infants (Bard 1994; Bard et al. 2005). Mothers give food to infants either actively or passively (Ueno and Matsuzawa 2004), and infants refer to their mothers before mouthing or ingesting novel foods (Ueno and Matsuzawa 2005). Infant chimpanzees also learn to use tools by observing their mothers at a close distance (Hirata 2006b; Hirata and Celli 2003; Matsuzawa et al. 2001).

The examples described in this chapter tell us that the core maternal behaviors including holding and suckling an infant are not always possible for chimpanzee mothers. They need appropriate environments, experience, and learning before becoming able to accomplish these behaviors. At the same time, the foregoing examples show how chimpanzee mothers can solve situations by their own means. These examples are not exceptional among chimpanzees. Bard (1994) described interactional caregiving contexts observed in captive chimpanzees. Included in these contexts were safeguarding in which the mother protected her infant from harm, soothing by which a mother acted to calm the infant, exercising in which the mother encouraged motor development, and assessment in which the mother checked the infant's muscle tone or behavioral state. From evolutionary and developmental perspectives, the long periods of infant dependency on parents provides an essential foundation for the development of the cooperative and altruistic aspects of social intelligence.

Acknowledgments. These studies were financially supported by the Japan Society for the Promotion of Science (JSPS) for Young Scientists (No. 2926) to S. Hirata, the Core-to-Core Program HOPE by JSPS, and Grants-in-Aid for Special Promotion Research (#12002009, #16002001) of the Ministry of Education, Culture, Sports, Science and Technology to T. Matsuzawa. We greatly acknowledge T. Matsuzawa for his generous guidance throughout the studies, M. Tomonaga and M. Tanaka for their support and suggestions, and all the collaborators at the Primate Research Institute of Kyoto University and Hayashibara Great Ape Research Institute for their assistance in data collection and analysis.

References

Bard KA (1992) Intentional behavior and intentional communication in young free-ranging orangutans. Child Dev 63:1186–1197

Bard KA (1994) Evolutionary roots of intuitive parenting: maternal competence in chimpanzees. Early Dev Parenting 3:19–28

Bard KA (2000) Crying in infant primates: insights into the development of crying in chimpanzees. In: Barr RG, Hopkins B, Green JA (eds) Crying as a sign, a symptom, and a signal: clinical emotional and developmental aspects of infant and toddler crying. Cambridge University Press, New York, pp 157–175

Bard KA, Myowa-Yamakoshi M, Tomonaga M, Tanaka M, Quinn J, Costall A, Matsuzawa T (2005) Group differences in the mutual gaze of chimpanzees (*Pan troglodytes*). Dev Psychol 41(4):616–624

Bloomsmith MA, Kuhar C, Baker K, Lambeth S, Brent L, Ross SR, Fritz J (2003) Primiparous chimpanzee mothers: behavior and success in a short-term assessment of infant rearing. Appl Anim Behav Sci 84:235–250

Brent L, Williams-Blangero S, Stone AM (1996) Evaluation of the chimpanzee breeding program at the Southwest Foundation for Biomedical Research. Lab Anim Sci 46:405–409

Byrne RW, Whiten A (1988) Machiavellian intelligence: social expertise and the evolution of intellect in monkeys, apes, and humans. Clarendon Press, Oxford

Call J (2001) Chimpanzee social cognition. Trends Cognit Sci 5:388–393

Call J, Carpenter M (2001) Do apes and children know what they have seen? Anim Cognit 4:207–220

Davis D, Fouts RS, Hannum ME (1981) The maternal behavior of a home-reared, language-using chimpanzee. Primates 22:570–573

de Waal FBM (1982) Chimpanzee politics. Cape, London

Hare B (2001) Can competitive paradigms increase the validity of experiments on primate social cognition. Anim Cognit 4:269–280

Hare B, Call J, Agnetta B, Tomasello M (2000) Chimpanzees know what conspecifics do and do not see. Anim Behav 59:771–785

Hare B, Call J, Tomasello M (2001) Do chimpanzees know what conspecifics know? Anim Behav 61:139–151

Hiraiwa-Hasegawa M (1990) Maternal investment before weaning. In: Nishida T (ed) The chimpanzees of Mahale Mountains. University of Tokyo Press, Tokyo, Japan, pp 257–266

Hirata S (2006a) Tactical deception and understanding of others in chimpanzees. In: Matsuzawa T, Toimonaga M, Tanaka M (eds) Cognitive development in chimpanzees. Springer, Tokyo, pp 265–276

Hirata S (2006b) Learning and transmission of tool use to fish for honey in chimpanzees. In: Matsuzawa T, Toimonaga M, Tanaka M (eds) Cognitive development in chimpanzees. Springer, Tokyo, pp 201–213

Hirata S, Celli ML (2003) Role of mothers in the acquisition of tool-use behaviours by captive infant chimpanzees. Anim Cognit 6:235–244

Hirata S, Matsuzawa T (2001) Tactics to obtain a hidden food item in chimpanzee pairs (*Pan troglodytes*). Anim Cognit 4:285–295

Horvat JR, Kraemer HC (1981) Infant socialization and maternal influence in chimpanzees. Folia Primatol 36:99–110

Horvat JR, Coe CL, Levine S (1980) Infant development and maternal behavior in captive chimpanzees. In: Bell RW, Smotherman WP (eds) Maternal influences and early behavior. Spectrum, New York, pp 285–309

Humphrey NK (1976) The social function of intellect. In: Bateson PPG, Hinde RA (eds) Growing points in ethology. Cambridge University Press, Cambridge, pp 451–479

Matsuzawa T (2003) The Ai project: historical and ecological contexts. Anim Cognit 6:199–211

Matsuzawa T (2006a) Evolutionary origins of human mother-infant relationship. In: Matsuzawa T, Toimonaga M, Tanaka M (eds) Cognitive development in chimpanzees. Springer, Tokyo, pp 127–141

Matsuzawa T (2006b) Sociocognitive development in chimpanzees: a systhesis of laboratory work and fieldwork. In: Matsuzawa T, Toimonaga M, Tanaka M (eds) Cognitive development in chimpanzees. Springer, Tokyo, pp 3–33

Matsuzawa T, Biro D, Humle T, Inoue-Nakamura N, Tonooka R, Yamakoshi G (2001) Emergence of culture in wild chimpanzees: education by master-apprenticeship. In: Matsuzawa T (ed) Primate origins of human cognition and behavior. Springer, Tokyo, pp 557–574

Matsuzawa T, Nakamura M (2004) Caregiving: mother–infant relations in chimpanzees. In: Bekoff M (ed) Encyclopedia of animal behavior. Greenwood Press, Westport, pp 196–203

Mizuno Y, Takeshita H, Matsuzawa T (2006) Behavior of infant chimpanzees during the night in the first 4 months of life: smiling and suckling in relation to behavioral state. Infancy 915–234

Nankivell B, Fritz J, Nash L, Fritz P (1988) Competent maternal behavior by a hand-reared, resocialized, primiparous chimpanzee: a case history. Lab Primate Newsl 27: 5–8

Nicolson NA (1977) A comparison of early behavioral development in wild and captive chimpanzees. In: Chevalier-Skolnikoff S, Poirier FE (eds) Primate bio-social development: biological, social, and ecological determinants. Garland, New York, pp 529–560

Nishida T, Hosaka K (1996) Coalition strategies among adult male chimpanzees of the Mahale Mountains, Tanzania. In: McGrew WC, Marchant LF, Nishida T (eds) Great ape societies. Cambridge University Press, Cambridge, pp 114–134

Ochiai T, Matsuzawa T (1997) Planting trees in an outdoor compound of chimpanzees for an enriched environment. In: Hare V (ed) Proceedings of the third international conference on environmental enrichment. The Shape of Enrichment, San Diego, CA, pp 355–364

Parker ST, McKinney ML (1999) Origins of intelligence: the evolution of cognitive development in monkeys, apes, and humans. Johns Hopkins University Press, Bethesda

Premack D, Woodruff G (1978) Does the chimpanzee have a theory of mind? Behav Brain Sci 4:515–526

Purvis A, Webster AJ, Agapow P, Jones KE, Isaac NJB (2003) Primate life histories and phylogeny. In: Kappeler PM, Pereira ME (eds) Primate life histories and socioecology. University of Chicago Press, Chicago, pp 25–40

Rogers CM, Davenport RK (1970) Chimpanzee maternal behavior. In: Bourne GM (ed) The chimpanzee, vol 3. Immunology, infections, hormones, anatomy, and behavior of chimpanzees. University Park Press, Baltimore, pp 361–368

Ross C (2003) Life history, infant care strategies, and brain size in primates. In: Kappeler PM, Pereira ME (eds) Primate life histories and socioecology. University of Chicago Press, Chicago, pp 266–284

Struthers EJ, Bloomsmith MA, Alford PL (1990) A case history of a decrement in maternal competence in a captive chimpanzee (*Pan troglodytes*). Lab Primate Newsl 29:3–6

Takemoto H (2004) Seasonal change in terrestriality of chimpanzees in relation to microclimate in the tropical forest. Am J Phys Anthropol 124:81–92

Takeshita H, Myowa-Yamakoshi M, Hirata S (2006) A new comparative perspective on prenatal motor behaviors: preliminary research with four-dimensional (4D) ultrasonography. In: Matsuzawa T, Toimonaga M, Tanaka M (eds) Cognitive development in chimpanzees. Springer, Tokyo, pp 37–47

Tomasello M, Call J, Hare B (2003a) Chimpanzee understand psychological states: the question is which ones and to what extent. Trends Cognit Sci 7:153–156

Tomasello M, Call J, Hare B (2003b) Chimpanzee versus humans—it's not that simple. Trends Cognit Sci 7:239–240

Tomonaga M, Tanaka M, Matsuzawa T, Myowa-Yamakoshi M, Kosugi D, Mizuno Y, Okamoto S, Yamaguchi M, Bard K (2004) Development of social cognition in infant chimpanzees (*Pan troglodytes*): face recognition smiling gaze and the lack of triadic interactions. Jpn Psychol Res 46:227–235

Ueno A, Matsuzawa T (2004) Food transfer between chimpanzee mothers and their infants. Primates 45:231–239

Ueno A, Matsuzawa T (2005) Response to novel food in infant chimpanzees: do infants refer to mothers before ingesting food on their own? Behav Proc 68:85–90

van de Rijt-Plooij HHC, Plooij FX (1987) Growing independence, conflict and learning in mother–infant relations in free-ranging chimpanzees. Behavior 101:1–86

Whiten A (1999) Parental encouragement in gorilla in comparative perspective: implications for social cognition and the evolution of teaching. In: Parker ST, Mitchell RW, Miles HL (eds) The mentalities of gorillas and orangutans. Cambridge University Press, Cambridge, pp 342–366

Whiten A, Byrne RW (1988) Tactical deception in primates. Behav Brain Sci 11: 233–273

Whiten A, Byrne RW (1997) Machiavellian intelligence. II: Extensions and evaluations. Cambridge University Press, Cambridge

3
Primates' Use of Others' Gaze

James R. Anderson and Sarah-Jane Vick

1. Introduction

For humans and many other animals, being gazed at can be a powerful stimulus. In many species too, one individual gazing toward something in the environment may influence where another individual looks; gaze direction can potentially convey much information. Observational and experimental studies have aimed to throw light on the contexts and functions of this kind of information gathering, the neural substrates underlying gaze perception, impairments to its effectiveness, and the role of different cues in the transfer of information. In humans, early gaze-tracking leads to joint attention and ultimately to hypotheses about other individuals' emotional and cognitive states. In nonhuman primates, responses to gaze cues from other individuals suggest the evolution of a mosaic of abilities and sensitivities, some of which are explored in the present chapter.

The challenges of surviving in complex social environments may have largely driven the evolution of primate cognitive abilities (Barrett et al. 2003; Byrne and Whiten 1988; Humphrey 1976). The evolution of group living has also been accompanied by a progressive increase in visually based communication (Andrew 1963; Brothers 1996; Emery 2000) and important information about the social and ecological environments is obtained by visually monitoring group mates (Kummer 1967). Attending to faces is central to social information gathering, for extracting categorical information such as identity or sex (Parr and de Waal 1999; Pascalis et al. 1999; Perrett and Mistlin 1990; Preuschoft 2000), more transient information such as hormonal status (Roberts et al. 2004; Setchell and Dixson 2001), and dynamic information in communicative facial displays (Darwin 1872; Ekman and Rosenberg 2005; van Hooff 1967). The eyes are of particular interest as gaze is uniquely both a signal and channel; while an individual is gathering visual information about the environment it is also signalling this attention to others by virtue of its visual orientation (Argyle 1988). How primates, including humans, interpret this language of the eyes has become an important question in the study of social cognition. As comparative developmental psychologists, we are interested in the ontogeny and phylogeny of these sophisticated gaze-reading abilities.

Department of Psychology, University of Stirling, Stirling, FK9 4LA Scotland, UK

2. Gaze Processing in Human Adults

Gaze in a basic component of human interaction. We are extremely sensitive to being looked at (Anstis et al. 1969; Gibson and Pick 1963). In a visual search task, direct gaze target faces were detected faster than averted gaze face targets and other stimuli (von Grünau and Anston 1995). Changes around the eye region play a major role in facial expression interpretation, for example during smiling (Frank et al. 1993; Seyama and Nagayama 2002). Patterns of gaze regulate turn-taking during conversation (Argyle and Cook 1976; Kleinke 1986). Furthermore, gaze direction mediates the interpretation of expressions, in terms of whether these are directed towards the perceiver, particularly for complex social emotions such as shame and embarrassment (Keltner 1995). Gaze cues may also provide a window onto a range of less emotionally charged cognitive states, as illustrated by the Reading-The-Mind-In-The-Eyes test (Baron-Cohen et al. 2001). For example, normal human adults show good agreement when asked to judge an individual's cognitive state (e.g. pensive or decisive) as well as feeling (e.g. embarrassed or annoyed) when only a photograph of the eye region is seen. Gaze direction may signal "thinking" (or cognitive load) in both adults (Glenberg et al. 1998) and children (Doherty-Sneddon et al. 2002), with gaze aversion correlating with response difficulty. Perceived gaze direction also impacts upon the evaluation of others, including how attractive they appear (Kampe et al. 2001). For example, a recent study found that a female's gaze shift towards the viewer increases men's ratings of both her likeability and attractiveness, but increases ratings only of the former by women (Mason et al. 2005). On a variety of levels, gaze and its concomitant perception impacts considerably upon our everyday social interactions.

A fundamental aspect of gaze processing concerns detecting direct or mutual gaze from gaze directed elsewhere, as significant interactions often follow establishment of eye contact. However, gaze directed away from the perceiver can also lead to changes in attention in the latter. There is now a voluminous literature on these responses to shifts in attention in others, most notably in terms of how this ability develops and what psychological mechanisms underlie the responses. The phenomenon of visual orientation with others seems to involve rapid, automatic processing. Thus, central gaze stimuli have been found to affect performance in target detection tasks; despite being non-predictive of target location, gaze cues reliably cause a shift of attention, so that gaze-congruent targets are detected more rapidly that incongruent targets (Friesen and Kingstone 1998; Langton et al. 2000). Similar effects are found for other social cues; for example, head orientation cues enhance target detection, but only when bodily orientation is towards the perceiver (Hietanen 2002). This suggests that bodily, head and eye direction are integrated during processing (Perrett and Emery 1994) and that we are particularly sensitive to the gaze shifts of individuals who are oriented towards us.

While such processing of gaze cues was initially considered unique, recent studies suggest that similar responses can also be elicited using other stimuli, such as arrows (e.g., Ristic et al. 2002). However, at the neuropsychological level gaze does appear to receive specific processing; fMRI responses differed systematically according to whether ambiguous central stimuli were labelled as eyes or a vehicle (Kingstone et al. 2004). The attention shift-cueing paradigm has been used to study clinical populations. Split-brain patients only demonstrate reflexive gaze-following in the hemisphere specialised for face processing (Kingstone et al. 2000), again suggesting that gaze is a distinct percept. Although lack of sensitivity to gaze is often seen as characteristic in autism (Baron-Cohen 1995), autistic children demonstrate normal patterns of reflexive gaze-following (Swettenham et al. 2003). However, autistic children were reported to respond equally to gaze cues and arrows, while control comparisons with developmentally normal children showed preferential responding to the social cue (Senju et al. 2004). It may be that it is an inability to sustain joint attention that underlies some of the social difficulties evident in autism (Baron-Cohen et al. 2001).

3. Gaze Processing in Human Infants

Given the known complexities of gaze-reading in adults, the development of these abilities has been the focus of numerous developmental studies. Human neonates are sensitive to direct gaze when tested using a habituation paradigm, and they prefer a face that is looking towards them (Batki et al. 2000; Farroni et al. 2002). The ability to follow another's gaze is crucial in the development of social interactions (Werner and Kaplan 1963). Starting with Scaife and Bruner (1975), several studies have mapped the development of the tendency to co-orient visually with another person whose direction of attention changes. There is some debate regarding the emergence of this ability, perhaps due to methodological differences across studies in terms of both independent (e.g. head or eye cues) and dependent variables (see Corkum and Moore 1995). Beyond these first signs of gaze-following, more sophisticated abilities progressively emerge. Butterworth and Jarret (1991) identified three stages of gaze-following: the ability to follow gaze in a general direction from around 6 months, the ability to follow gaze geometrically and detect target of attention (and ignore distractors) at around 12 months, and to follow gaze outside one's own visual field (representational gaze-following) at around 18 months.

Although gaze-following by young infants was originally shown using congruent shifts of head and eyes, it has been shown that shifts in eye direction alone may trigger co-orientation (Hood et al. 1998). By the age of 18 months human infants tested with a live model can also respond to eye gaze alone; infants at this age show geometric gaze-following, and take into account visual barriers and locations outside their own visual field. They are also likely to check back if they find nothing of interest upon following another's gaze, as if seeking further information.

It is not evident that at this young age children have a mentalistic understanding of gaze; this capacity may not emerge until the age of 3 years, when children can use gaze to predict subsequent actions (Montgomery et al. 1998) and to infer knowledge states (Baron-Cohen 1994). By the age of 4 years children will use averted eye gaze to infer the act of "thinking", and visual fixation to identify desire or intention (as in the "sweets task", Baron-Cohen 1995). Moreover, attentional cues combine with emotional ones in various social contexts, including social referencing in which infants can learn the important features of their environment from significant others (e.g. Walden and Ogan 1988).

4. Gaze Processing in Nonhuman Primates

Although this chapter is restricted to primates, it should be noted that sensitivity to the presence or absence of eye-like stimuli is widespread in the animal kingdom (e.g., sparrows, Hampton 1994; plovers, Ristau 1998; snakes, Burghardt 1991; chickens; Gallup 1972). In primates, during face recognition tasks the eye region is particularly important (e.g., rhesus monkeys: Keating and Keating 1993; chimpanzees: Parr et al. 2000). When visually inspecting faces, rhesus monkeys pay special attention to the eyes whether the stimulus presented is a conspecific, chimpanzee, human or even only a schematic face (Gothard et al. 2004; Guo et al. 2003; Keating and Keating 1982; Sato and Nakamura 2001). In addition, primates are sensitive to the direction of the eyes and appear to find direct gaze perceptually compelling. Rhesus monkeys show increased interest in the eye region of pictures showing human and conspecific faces with direct gaze (Keating and Keating 1982; Kyes and Candland 1987; Nahm et al. 1997; Sato and Nakamura 2001). In a forced choice discrimination task with paired photographs of a human model, rhesus monkeys can discriminate direct gaze from gaze averted by as little as 5 degrees laterally (Campbell et al. 1990; Eacott et al. 1993). The most important things about eyes may well be whether or not they are looking at you; direct gaze seems to be a distinct percept which receives rapid processing (Kummer et al. 1996).

Distinctive behavioural responses to direct gaze are reported in several species of macaques (e.g., Exline and Yellin 1969; Kummer et al. 1996; Perrett and Mistlin 1990) and lesser mouse lemurs (Coss 1978), among others. In a study of 6 species of monkeys (talapoin, patas, longtailed, rhesus, stumptailed, squirrel), species, sex and age differences were found in terms of readiness to engage in eye contact with a human (Thomsen 1974), though these were not considered from the viewpoint of previous experience with humans. Also, only frequency of eye contact was recorded. The absence of any information about duration or other behavioural reponses makes it difficult to interpret these differences; for example, frequent eye contact could suggest monitoring, whereas prolonged eye contact might be indicative of affiliative or agonistic gestures (Emery 2000).

5. Emergence of Visual Co-orientation

A tendency to visually co-orient could become increasingly adaptive as young individuals become more independent and thus vulnerable to predation (Povinelli and Giambrone 2000; Tomasello et al. 2001). As in humans, nonhuman primate sensitivity to direct gaze is present from early infancy. By three weeks, rhesus macaque infants viewing conspecifics foveate more and show more emotional arousal when viewing images presenting direct eye contact compared to averted gaze (Mendelson et al. 1982). When confronted with an observer, young rhesus infants (9–12 weeks, Kalin et al. 1991) and older infants (6–12 months, Kalin and Shelton 1989) showed differential behavioural responses to gaze direction, though this included head orientation. An infant chimpanzee was shown to prefer pictures of a human displaying direct gaze (Myowa-Yamakoshi et al. 2003), but the preference disappeared when the facial features were scrambled.

The ability to follow the gaze of a human has now been studied in a range of primate species, starting with Itakura (1996), who tested two species of lemurs, four species of macaques, squirrel monkeys, two species of capuchin monkeys, chimpanzees and a human-raised orangutan. After establishing eye contact with the subject, the human presented combined head and eye turns either with or without accompanying pointing. When gaze shifts were accompanied by a pointing gesture, monkeys sometimes re-oriented, but only the apes did so reliably. Even without the manual gesture the chimpanzees oriented in the correct direction on 20% of trials, the orangutan on 70%. However, as visual co-orientation was related to amount of interest in the human, the results may not be an accurate reflection of gaze-following abilities (Itakura 1996). In a study comparing stumptailed macaques and black lemurs, only the macaques showed visual co-orientation in response to a human's switch of attention signalled by contiguous head and eye movements (Anderson and Mitchell 1999.) On some trials the macaques displayed delayed co-orientation, waiting more than 2 seconds before following the human's gaze shift, and clearly looking at the human's face during this period. Olive baboons also changed visual orientation in response to a shift in a human's head direction, even when the latter's eyes were closed. The same baboons also tended, albeit less reliably to follow a shift in eye gaze only (Vick & Anderson, unpublished). Although young pigtailed macaques (2–6 years old) co-oriented to congruent head and eye shifts by a human, only adults responded reliably to eye gaze alone (Ferrari et al. 2000).

Gaze-following has been studied mostly in chimpanzees, who co-orient upon seeing a "distracted" human (Povinelli and Eddy 1996a) and attend to the same quadrant of space, even following shifts in eye direction alone (Povinelli and Eddy 1996c, 1997). The ability to follow another's gaze to a location outside one's own visual field is considered significant because human infants do not do this until they reach about 18 months of age, which some observers interpret as an important landmark in perspective-taking (Butterworth 1995; Moore 1999; but see Tomasello 1995). Bräuer et al. (2005) found no significant differences in gaze-following responses across all the species of great apes, but there were differences between age groups. Adults showed a greater tendency to co-orient with a human and to check back when they did not locate any target of gaze. Okamoto et al. (2002) showed that, following training, an infant chimpanzee was able to follow a human's gaze by around 13 months, but more naturalistic studies have reported a slightly later onset of this ability, at around 2–3 years (Brauer et al. 2005;

Tomasello et al. 2001). Methodological differences may account for divergent findings; the 13-month-old was tested on gaze-following to objects already within its visual field, whereas other studies have assessed gaze-tracking to a more distal and often targetless location in space.

Using cross-sectional and longitudinal data, Tomasello et al. (2001) reported that rhesus macaques and chimpanzees both developed the ability to co-orient with a human who switched head and eye orientation during mid-to-late infancy (5.5 months and 3–4 years, respectively). Only older individuals (rhesus from 2 years of age and chimpanzees from 4 years) learned to ignore uninformative cues; that is, when presented with repeated trials they became less likely to co-orient (Tomasello et al. 2001). Nonhuman primates may start to visually co-orient with others at an early age, but with increasing maturity the behaviour becomes more flexible as they learn to control the co-orienting response.

Although the emergence of spontaneous gaze-following has been explored in pigtailed and rhesus macaques (Ferrari et al. 2000) all species of great apes (Bräuer et al. 2005; Tomasello et al. 2001) and a gibbon (Myowa-Yamakoshi and Tomonaga 2001), it would be valuable to have further information on more species. One question that could drive new work in this area is the extent to which the pattern and time course of gaze-following/visual co-orientation is comparable to that observed in human infants. For example, human infants tend to respond to head orientation rather than eye gaze until they reach around 18 months (Moore 1999). The results with nonhuman primates are as yet unclear. For example, the broad age range (2–6 years) used by Ferrari et al. (2000) does not give much indication of when this ability emerges in macaques. Tomasello et al. (2001) suggest that when chimpanzee gaze-following emerges as a reliable response, it may already be at a sophisticated level. In that study, chimpanzees who reliably visually co-oriented had also performed well in an earlier study examining their abilities to accurately locate the target of another's gaze (Tomasello et al. 1999).

One recent study looked at monkeys' responses to different combinations of attention shifts by humans (all previous studies had only one model). Anderson et al. (2005) found that capuchin monkeys showed overall more co-orientation than squirrel monkeys, with some suggestion that the two species respond differently to different combined attention events, for example one person switching attention twice, compared to two people switching attention either simultaneously or consecutively. This raises an interesting question: Are group-living primates more sensitive to attention shifts by more than one conspecific? We also know little about how humans deal with multiple attention shifts (see Anderson and Doherty 1997 for examples).

6. Visual Co-orientation to Conspecifics and 2D Stimuli

Visual co-orienting must have evolved on the basis of countless situations in which responding this way proved advantageous. In a experiment involving the sudden appearance of a food item, five species of primates co-oriented with the group mate that saw and reacted visually to the food (chimpanzees, sooty mangabeys, rhesus, stumptailed and pigtailed macaques, Tomasello et al. (1998)). In all five species, co-orientation usually occurred within 1 second. In contrast, no such response occurred in control trials in which no food appeared. However, it is not clear what role cues other than eye gaze (bodily orientation, head direction) might have played in eliciting these co-orientation responses.

The cues used by monkeys when co-orienting have been examined using schematic and photographic images as eliciting stimuli (e.g., Lorincz et al. 1999). Two rhesus monkeys responded more to head orientation compared to body posture when these conflicted, and although they responded to eye gaze alone, they were more likely to respond when head and eye orientation were congruent. Ferrari et al. (2000) studied the scan patterns of two head-restrained adult pigtailed macaques presented with various gaze signals shown in photographs of a human (head + eyes, eyes; also a non-social control condition). The monkeys reliably followed the human's gaze to look at the same quadrant of space in both the head + eyes and eyes alone conditions, but not during the control condition.

Fagot and Deruelle (2001) used a cueing paradigm, based on one often used to demonstrate gaze-induced reflexive orienting in humans (e.g. Friesen and Kingstone 1998), to examine baboons' responses to gaze. Schematic and photographic faces depicting eye gaze were presented in the centre of a monitor and, following a brief interval. (300 ms), a target would appear either on the side congruent with the depicted eye gaze or on the opposite side of the monitor. In humans, responses to congruent targets (whether detection, localisation or identification of targets) are significantly faster than to incongruent targets. In contrast, the baboons' response times did not differentiate between cued and uncued targets (Fagot and Deruelle 2001). However, after extensive exposure to a condition in which gaze consistently predicted target location, the baboons became faster to respond to congruent than incongruent targets; that is, they learned to use the depicted eye direction to locate targets.

Deaner and Platt (2003) tested humans and rhesus monkeys for reflexive gaze-following to conspecific models and found that non-predictive head and eye-gaze cues induced gaze shifts to peripheral targets. It is unclear why Fagot and Deruelle (2001) failed to obtain reflexive responses to gaze in baboons whereas Deaner and Platt (2003) succeeded with macaques (see also Shepherd et al. 2006). Conceivably, differing task demands and dependent variables may be important; the former study required manual responses to centrally cued targets whereas the latter measured eye movements in response to centrally presented stimuli.

7. The Object of the Other's Attention?

Although many primates show gaze-following, the question remains whether they use gaze to anticipate and explain others' actions (Gordon 1998). While a general tendency to co-orient may be advantageous, accurately detecting the specific target of another's attention would be even more so. The ability of young chimpanzees to understand another's attention was tested with barriers that intersected a human's line of regard (Povinelli and Eddy 1996c); when the human glanced toward the partition, the chimpanzees responded by attempting to examine the partition and not the wall behind themselves. A recent study involving barriers found that orangutans, chimpanzees, gorillas and bonobos all moved to try to gain visual access to the area a human attended to (head and eye direction combined), although their visual attention was not directly measured. They also remained longer in this location compared to control trials in which the human looked up towards the ceiling (Bräuer et al. 2005). When chimpanzees were tested with both barriers and distractor objects (Tomasello et al. 1999), the apes looked around the barriers and both at and beyond the distractor objects when these locations were fixated by the trainer. The chimpanzees therefore demonstrated an ability to follow gaze "geometrically" and accurately project another's line of sight (see Butterworth and Jarret 1991).

Emery et al. (1997) used conspecific images presented on video to examine rhesus monkeys' responses to another's target-directed gaze. The eye movements of two monkeys were recorded as they viewed an image of a conspecific looking at a target; the eye and head direction and body posture were all oriented towards the target object. The monkeys fixated a target object that the model monkey was oriented towards more than an identical distractor object, and the authors stated that they had identified an ability to follow gaze and engage in joint visual attention. However, the monkeys also fixated the target location before any objects appeared, and they failed to maintain interest in the object once the monkey's image was removed. Thus, the results reflect an ability to follow gaze, but it is not clear whether joint attention also occurred (see Emery 2000). Scerif et al. (2004) reported that Diana monkeys responded to pictorial gaze cues in terms of duration and first inspections to target side. Again, interest was higher prior to the appearance of the objects, indicating a gaze-following response to a spatial location rather than joint visual attention directed at a common object of interest. Baboons tested for their ability to categorise target-directed gaze depicted in photographs of humans required extensive training to master the task to a criterion of 80% correct trials (Vick et al. 2001). Furthermore, any knowledge about gaze that was picked up during the photographic discrimination training did not transfer to a situation involving a live human presenting gaze cues in an object-choice task (see below). The findings suggested the use of simple perceptual cues to perform discriminations, rather than understanding another's attention to an object.

8. Gaze Cues and Object-choice Tasks

In the typical object-choice task, the subject is offered experimenter-given behavioural cues, such as gaze or manual gestures, that can be used to locate a hidden food item in, under or behind one of two objects (Anderson et al. 1995). Note that simple co-orientation alone should heighten the probability of choosing the first object encountered and retrieving the food item. However, despite their tendency to follow gaze, most primates tested encounter difficulties on object-choice tasks (Anderson et al. 1995; 1996; Call et al. 2000; Hare 2001; Vick and Anderson 2000). An early study with capuchin monkeys found that a human's pointing, not gaze, was necessary and sufficient for the monkeys to master the task. Despite over 1000 trials with head + eye orientation available as a cue, the monkeys did not use this source of information (Anderson et al. 1995). Cotton top tamarins showed some evidence of sensitivity to manual gestures but not gaze as experimenter-given cues (Neiworth et al. 2002). Similar results were reported for rhesus monkeys; none of the monkeys responded correctly to head and eye cues but two of the three showed improved performance with manual gestural cues (Anderson et al. 1996). One monkey did perform above baseline when a head and eyes cue was presented close to the baited object, but eye gaze alone was not exploited as a cue. Correction trials, time outs, and reduced cue-target distance were used with a capuchin monkey; the latter mastered (to 80% correct) manual (tapping and pointing) and head orientation cues, but not eye gaze (Itakura and Anderson 1996). When this method was used with other capuchins, two of the three monkeys were able to use eye gaze as a cue, but they favoured head direction when this conflicted with eye gaze (e.g., head oriented to the left, eyes to the right). Furthermore, performance diminished when another human presented the cues (Vick and Anderson 2000).

Itakura and Tanaka (1998) reported that great apes (two chimpanzees and an orangutan) and human children (aged 2 years) responded well to all cues up to and including eye gaze, although performance in the latter condition was lower than in the preceeding ones. Byrnit (2004) tested 3 orangutans and found superior performance with pointing than gaze as an experimenter-given cue, although two did show some evidence of gaze-reading. Gorillas were also able to use various experimenter-given cues, but they did not cooperate during eye gaze only trials (Peignot and Anderson 1999). However, both gorillas (Peignot and Anderson 1999) and orangutans (Byrnit 2004) had considerably less experience of interacting with humans compared to many of the other apes that feature in these types of study. Rearing histories and experience with humans are always likely to be important in nonhuman primates' responses to human trainers during cognitive testing (e.g. Call et al. 1998; Tomasello and Call 1997). For example, in the first study of gaze-monitoring in gibbons, a 3-year-old, human-reared gibbon was able to use pointing, head orientation and eye gaze to choose a baited object (Inoue et al. 2004).

Chimpanzees previously shown to follow gaze in response to a shift in a human's head and eye orientation used the same information less reliably in an object-choice task (Call et al. 1998). Furthermore, the type of container influenced performance. This study shows that even small procedural details may have important implications for performance in object choice tasks. Call et al. (2000) examined the type of cues from a human that chimpanzees would use to find hidden food items. Vocalizations and other noises, or the human approaching, touching or lifting and looking under the container, all facilitated performance in some chimpanzees, but vocalizations alone did not. Performance was also diminished when the cue consisted of eye direction alone. Itakura et al. (1999) reported that three out of twelve chimpanzees performed above chance given combined head + eye direction cues, whereas six showed enhanced performance when gaze was accompanied by vocalisations. Approach and vocal cues may assist the chimpanzees by engaging them in a "foraging mode" and may change the functional context of the task (Call et al. 2000); perhaps gaze-following relates to social monitoring, while locating food requires additional cues (Itakura et al. 1999). In any case, there seems to be something about the typical object-choice task that makes it particularly challenging for many or most nonhuman primates. Call et al. (2000) suggest that gaze-following may be a simple mechanism that facilitates scanning the environment for information, whereas the object-choice task is a communicative situation in which the relevance of the model's behaviour must be understood. Thus, the superiority of cue reading by nonhuman primates reared in extensive contact with humans may be due to their greater facility with human signals (Tomasello and Call 1997).

In contrast to evidence that chimpanzees can accurately follow another's gaze (Tomasello et al. 1999), Povinelli et al. (1999) reported that chimpanzees were insensitive to the attention of a human during an object-choice task. The human fixated the baited container using eyes only, or else oriented head and eyes together either towards or above the container; all cue conditions were presented both statically and dynamically. These chimpanzees had previously demonstrated sensitivity to eye gaze by visually co-orientating (Povinelli and Eddy 1996b), but the same cue resulted in only chance performance on the object-choice task. The authors suggest that responding to both on- and off-target gazes (head and eyes) indicates that the chimpanzees do not have a high-level appreciation of seeing (Povinelli et al. 1999). Clearly, the chimpanzees were not insensitive to gaze direction, as they co-oriented with the trainer's gaze during off-target trials. Although the trainer was not specifically cueing either container, head orientation in the general direction of one of them would at least seem to differentiate the two and indeed the chimpanzees tended to choose this container. However, in a comparative study with human children (3 years old), the children responded randomly when presented with such a "distracted" trainer, suggesting that the two species employ different strategies when faced with this ambiguous task (Povinelli et al. 1999).

As with general gaze-following, we know less about responses to cues from conspecifics in these tasks, but chimpanzees performed similarly when tested with both human and chimpanzee informants (Itakura et al. 1999). A local enhancement cue, in which the informant simply approached the baited object and remained there, was used by all four chimpanzees when given by a conspecific and by three when the actor was human. Only one chimpanzee successfully used a gaze (head + eyes) and a point cue, and only when these were presented by a human. However, the chimpanzee's initial performance was random, suggesting that the informational value of the gesture was learned during the study (Itakura et al. 1999).

9. Gaze in Competitive Contexts

The object choice task is essentially a cooperative interaction, albeit one-way, between cue-giver and responding individual. An interesting recent development has emerged from the recognition that competitive contexts may better reveal primates' abilities to exploit the gaze of others. Hiding from others may be one such example. Field observations indicate that this is a common form of "tactical deception" in nonhuman primates. For example, mating only when out of view of more dominant group-members, thereby avoiding aggression, may reveal something about the ability to judge another's visual orientation (Byrne and Whiten 1988). But when faced with a human who prevented their attempts to drink from a desired resource, captive longtailed macaques did not demonstrate any preference for concealement by an opaque partition (Kummer et al. 1996). In a more naturalistic setting, low- and middle-ranking male long-tailed macaques were more likely to mate near occluders than visually open areas (Gygax 1995); however, there was no evidence of any tendency to hide during aggression (Gygax 2000). Furthermore, there was no preference for concealment behind solid partitions over more poorly camouflaged locations (panels offering only partial concealment; Gygax 1995). Evidence for hiding abilities in monkeys is limited and, in any case, need not imply perspective-taking, as simpler mechanisms may explain the behaviour, for example simply avoiding another's direct gaze.

10. Competition Over Food

Chimpanzees appear able to appreciate a conspecific's line of regard in a competitive situation (Hare et al. 2000). Two chimpanzees were paired for competitions over two desirable food items placed within a communal area. Each chimpanzee was initially restricted to a room on opposite sides of the communal area. The food items were placed in various locations so that only the subordinate chimpanzee could see both items. The behaviours of both chimpanzees when given access to the area were recorded. The subordinates consistently chose the food item that they could see but their rival could not, rather than the one that both could see. In contrast, dominants would first secure the openly visible item before taking the one that they alone saw. This finding was robust across a number of manipulations, for example, when tyres or barriers were used to conceal food items. However, things changed when a transparent barrier was introduced, indicating that choices were not simply made on the basis of accessibility or proximity of food items to objects. Chimpanzees were tested both as dominant and subordinate members of dyads, and they showed flexible responses dependent upon the identity of their competitor, making simple rule learning about occluders an unlikely explanation of their behaviour (Hare et al. 2000). This facility for using a conspecific's visual orientation to choose between two available food items contrasts with the results of the object-choice study reported above (Itakura et al. 1999). Clearly, it is not the foraging situation alone that influences processing of cues about another's attention, but the context may also be crucial. (Hare 2001).

Chimpanzees readily exploit information from another's looking behaviour in a competitive food problem, but do so less naturally in the "cooperative" context of the typical object-choice task. A direct comparison between performance on cooperative and competitive tasks showed that chimpanzees were able to exploit manual gestures within both contexts, but spontaneously performed better when presented with a competitive human or conspecific than with a cooperative human (Hare and Tomasello 2004). It is interesting to note that capuchin monkeys tested using the competitive paradigm differed from chimpanzees in that while they successfully selected a concealed food item during conspecific competition, they did not do so when given a slight head start (Hare et al. 2003). This indicates that the capuchins were responding to behavioural cues from their competitor rather than showing visual perspective taking. When task complexity was reduced by presenting just a single visible or hidden food item, the capuchins still failed to demonstrate an appreciation of their competitior's visual access to the item.

In a variant of the competitive situation, baboons rapidly proved adept at exploiting a human's visual orientation in order to remove the food item that was *not* fixated (Vick and Anderson 2003). These baboons used both head and eye cues quickly and without any of the supplementary training procedures that are often used to establish object-choice performance with experimenter-given cues. Manual gestures were found to be more effective than head direction, which in turn was more effective than eye gaze; when head and eye cues conflicted, head direction was the preferred cue. Flombaum and Santos (2005) recently studied rhesus macaques in a competitive food situation involving a pair of humans. The monkeys preferentially took food from in front of a non-attending human, as indicated by bodily orientation, head direction, or eye direction alone. Moreover, the monkeys were able to take visual barriers into account, selecting the competitor whose gaze (both face or eyes alone) was not occluded by an opaque barrier. These studies suggest that competitive paradigms may be more appropriate for assessing nonhuman primate social cognition, at least those that involve access to food resources. However, we know little about gaze-monitoring in other contexts, such as mating situations, although evidence for tactical deception suggests that we should expect some degree of visual perspective-taking (Byrne and Whiten 1988; but see Gygax 1995).

11. Gaze and Intentions

Humans use others' looking behaviour to predict subsequent actions. That is, we interpret gaze as a signal of interest in, and intentions towards other individuals or objects (e.g. Montgomery et al. 1998). One approach to studying whether primates also interpret gaze as a signal of intention has been to use an expectancy violation paradigm, commonly used to study simple psychological processes in pre-linguistic human infants. Reaching for an item previously attended to can be seen as an expected act, whereas reaching for another, unattended item is not expected. Differential responses in these conditions, specifically, increased interest in unexpected sequences, could indicate an understanding of gaze as signal of intent. This is precisely what was reported in cotton-top tamarins. These small New World monkeys showed evidence of sensitivity to a human's combined head and eye orientation but not to eye gaze alone (Santos and Hauser 1999). They showed some capacity for predicting actions based on preceding visual orientation, although the mean increase in duration of looking response amounted to less than a 1-second, wtih results based on only one trial per monkey. Moreover, because the monkeys were exposed to three familiarisation trials in which the human looked at and manipulated the same object, it is possible that they simply learnt to associate the two events. Simple co-orientation could also cause human and monkey to look at the same object so that manipulation of the other object by the former causes surprise. Two other species of New World primates tested using similar procedures (capuchins and squirrel monkeys) did not show such clear-cut expectancy violation (Anderson et al. 2004). Following habituation to congruent gaze (head and eye direction) and object manipulation sequences, neither species showed expectancy violation when presented with incongruent sequences. In addition, the omission of the familiarisation trials did not reveal any tendency to respond differentially to congruent and incongruent gaze-action sequences. We need more information on how other primate species might respond in these kinds of expectancy violation experiments.

Staying with the expectancy violation, it is interesting to consider another phenomenon observed during gaze-following, namely checking back. Call et al. (1998) report that when gaze-following in response to a shift in head and eye orientation, chimpanzees sometimes look back to the human's face if they find nothing of interest, before co-orienting again. Bräuer et al. (2005) report that great apes followed shifts in a human's combined head and eye orientation, and checked back when they found no target of attention. Scerif et al. (2004) state that Diana monkeys checked back to the stimulus monkey photograph when a target appeared at a location incongruent with orientation cues. "Checking back" is considered an important facet of child socio-cognitive development, though there is some debate regarding its meaning. For some authors, it is indicative of a mentalistic appreciation of gaze, whereas for others simpler processes may underlie it (Corkum and Moore 1995). Call et al. (1998) considered the possibility that the chimpanzees simply returned to a central orientation once they had finished gaze-following, so that the second gaze tracking response was an independent occurrence (Call et al. 1998). However, more detailed analysis of response patterns reveals that checking back is not directly related to the initial tendency to co-orient. As checking back emerges during development, it seems dissociable from the basic co-orienting mechanism that is seen in younger apes (Bräuer et al. 2005). The emergence of looking back and habituation to gaze cues suggest that while younger primates may already be equipped with a simple co-orientation response, older individuals seem to show more understanding of others' gaze behaviours; this better understanding mediates their visual co-orientation (Bräuer et al. 2005).

Some studies have directly examined primates' understanding of the relationship between seeing and knowing. Although an early study had concluded that chimpanzees recognised the relationship between seeing and knowing (Povinelli et al. 1990), a critical review by Heyes (1998) suggested a re-evaluation of the findings. In the original procedure, chimpanzees could choose to follow a signal from a human who had seen an object being baited with food (Knower) or one who had not (Guesser). The chimpanzees showed a preference for the Knower when the Guesser was not in the room during baiting, but did not demonstrate an immediate transfer to a condition in which both humans remained in the room and the Guesser wore a paper bag over her head so as not to see the baiting procedure. The chimpanzees quickly learned to use this new information, but they did not discriminate between the humans immediately. It may be that the presence of two humans simply made the task overly complex (Povinelli and Eddy 1996a). Alternatively, perhaps chimpanzees do not understand seeing as an epistemic state, but rather use simpler rules. For example, after many trials chimpanzees could distinguish the Knower from Guesser in terms of recognising which individual was present when baiting occurred (Povinelli et al. 1990), but this does not require any appreciation of seeing per se. As the studies in this section highlight, designing appropriate tasks to examine whether nonhumans appreciate that seeing leads to knowing is indeed challenging (Heyes 1998). Adapting the Knower-Guesser procedure in a study with capuchin monkeys, Kuroshima et al. (2002) demonstrated that the monkeys gradually learned to choose a container manually indicated by a Knower, who looked into the containers before cueing, in preference to a Guesser who indicated a random container. In subsequent conditions, these monkeys adapted to variations in which the human indicated the correct container without looking, and Knower and Guesser alternated randomly across trials. While these results suggest that capuchin monkeys can learn to recognize the relationship between seeing and subsequent accuracy of cues, it does not indicate that these monkeys readily understand the relationship between seeing and knowing.

A study within the (possibly) more ecologically relevant competitive paradigm (Hare et al. 2001) demonstrated that chimpanzees are able to keep track of what competitors have previously seen. The chimpanzees responded differentially according to whether or not a dominant competitor had previously witnessed a food item being hidden or moved to a new location. The subordinate chimpanzees obtained significantly more food items when their competitor was uninformed or misinformed about the location of food items; they also responded appropriately when knowledgeable competitors were replaced by naïve dominant competitors, indicating that chimpanzees at least are able to understand the relationship between seeing and knowing.

In a study of responsiveness to conspecific attention, Cheney and Seyfarth (1991) showed that rhesus and Japanese macaques interacting with an infant were sensitive to whether or not the mother of the infant was visually monitoring the encounter. The monkeys were separated by either a glass partition, an opaque barrier, or a one-way mirror that allowed the monkey and infant to see the mother but not vice versa (an effect that the monkeys had previously learnt through exposure to the mirror). Measures of approaches, retreats and agonism revealed that interactions between monkeys and infants were not simply affected by the mother's presence (one-way mirror) but rather by whether the mother was able to perceive the interaction (glass condition). However, the lack of contingency in the mother's behaviour in the one-way mirror condition might have played as much of a role as an appreciation of her visual attention.

12. Understanding the Role of Gaze in Visual Communication

In humans, gaze plays a remarkably flexible role in communication, in which eye contact can be an ostensive behaviour, both expressing and assessing communicative intent. According to Gómez (1996), great apes also use eye contact in an ostensive manner, for example, to gain attention of humans and make requests. Appropriate visual orientation is necessary for gestural communication, and apes may use non-visual gestures to gain someone's attention (Gómez 1996; Russell et al. 2005; Tomasello et al. 1994).

In a series of studies on gaze comprehension (Povinelli and Eddy 1996a), young chimpanzees (aged 5–6 years) were required to determine whether or not they were within a human's visual field before gesturing. Each chimpanzee was presented with two trainers, one looking at the chimpanzee and another not visually oriented towards it. The chimpanzee chose which trainer to beg from. In one condition in which one trainer faced the chimpanzee while the other had his/her back turned, the chimpanzees displayed an immediate and consistent disposition to gesture to the trainer facing them. In contrast, in all other conditions the chimpanzees failed to demonstrate any appreciation of whether or not the trainer could see them, although over time they did learn to choose the attentive trainer. These chimpanzees also showed a preference for eye contact with a human (Povinelli and Eddy 1996b), although tests revealed that head orientation and head movements also influenced preferences, suggesting that there may be more to chimpanzees' attention monitoring than eye direction alone.

Recent work on gestural communication in primates is revealing a complex interplay between gaze and signalling. For example, captive chimpanzees use more visual gestures towards individuals already attending, with gesture sequences emerging when recipients fail to respond to initial signals (Liebal et al. 2004). In addition, rather than demonstrating effective attention-getting behaviours in other channels, such as vocalisations or tactile gestures, chimpanzees move position so as to signal within the visual field of recipients. This reiterates Menzel's (1974) much earlier suggestion that young chimpanzees try to gain another's visual attention when engaging in a food locating task.

In nonhuman primates that point, the gesture is often accompanied by looks towards the human's face, perhaps indicating sensitivity to the latter's visual orientation (Leavens et al. 1996; Woodruff and Premack 1979). Furthermore, apes used to interacting with humans gesture differentially as a function of whether or not a human's eyes are open (Call and Tomasello 1994; Gómez 1996). However, it is conceivable that the primates may simply learn that the visibility of a human's face, and perhaps even eye direction, mediate interactions (Povinelli and Eddy 1996a). In a study of chimpanzees' responses to an experimenter engaging in one of four behaviours (20 sec of either making and maintaining direct eye contact, making direct eye contact and "attentive" head movements, waiting with eyes closed, or looking above and behind the chimpanzee), the apes showed no differences in latency or amount of attention-getting behaviours exhibited (Theall and Povinelli 1999). Similarly, in what may be another demonstration of limited appreciation of another's ability to see; only one of four young chimpanzees removed a blindfold from a human who was helping to carry a container (Premack 1988). Overall, there is little evidence that primates use their attention in intentional ways, though some kinds of deception may be a sign of this, for example, attending to imaginary events, or feigning disinterest (Byrne and Whiten 1988).

One phenomenon considered important in the onset of intentional communication in human infants is gaze alternation (GA) during pointing (Bates et al. 1975). In GA the infant alternately shifts gaze between the pointed-at object and the partner's face; it is considered a hallmark of intentionality of the gesture. There is very little information on GA in nonhuman primates, although it has been described as accompanying gestures by great apes (e.g., Gómez 1996; Leavens). It also emerged in a macaque that engaged in eye contact with a human while pointing towards objects in her presence (Kumashiro et al. 2002). Recently, Anderson et al. (2007) reported GA in squirrel monkeys that were trained to point to a baited food container in the presence of a human. Sometimes the squirrel monkey GA involved combined head and eye movements, and sometimes it involved shifts in eye orientation alone. It was usually very fast. The discovery of GA in squirrel monkeys is interesting because these monkeys are considered to have "camouflage eyes" (see below; Kobayashi and Kohshima 2001).

Concerning use of gaze as indicator of intentionality, Tomasello (1995) suggests that nonhuman primates and human children below 1 year of age do not see others as intentional agents. Thus, even the gaze-following by infants should not be interpreted in a mentalistic manner (Butterworth and Jarret 1991); for young infants and nonhuman primates, gaze may be read only as directional cue and not as indicating a psychological state (Povinelli and Giambrone 2000). Tomasello (1995) further suggests that during their second year, human infants begin to appreciate intentionality and subsequently start to appreciate that looking behaviour can indicate attentional states; if nonhuman primates do not develop this understanding of intentionality then they cannot progress beyond a geometric appreciation of gaze. Clearly, our present knowledge of how primate socio-cognitive ability emerge and are shaped through experience is limited and further study is required to explore how gaze reading relates to understanding of intentions.

13. Species Differences

While primates including humans show a bias for monitoring the eye region of the face, the salience of the eyes themselves varies considerably across species. Human eyes are horizontally elongated and have the largest ratio of exposed sclera in primates; furthermore, this exposed sclera is devoid of pigmentation (Kobayashi and Kohshima 1997, 2001; Morris 1967). In the vast majority of other species of primates studied, the visible sclera is brown or dark brown in colour (Kobayashi and Kohshima 2001). The principle selection pressure upon eye colouration may be related to the communicative function of the eyes; visible sclera may render eye movements more visible to others (as in humans), while less salient coloration may serve to camouflage eye movements (Perrett and Mistlin 1990). For example, Thomsen (1974) noted that of several primate species studied, inter-observer reliability was lowest for squirrel monkeys, confirming a subjective impression that gaze is difficult to discern for small, dark eyes. For Kobayashi and Kohshima (2001), "camouflage eyes" that conceal eye direction could confer at least two advantages in nonhuman primates: avoiding conflict with conspecifics and avoiding predation.

There may be species differences in sensitivity to eye movements. In humans it is likely that the contrast between iris and white sclera is a key cue to gaze direction (Langton et al. 2000). For example, Ricciardelli et al. (2000) found that colour inversion confounded judgements of gaze direction. However, the psychophysics of gaze processing remains uncharted territory as far as other primates are concerned. In addition to scleral contrast, the timing (velocity and duration of eye movements, e.g. Blois-Heulin (1999)) and coordination of looking behaviours, such as head and eye direction (Kaplan and Rogers 2002) are all likely to impact upon responses to attentional states in others. In addition, use of gaze may vary systematically according to context, for example, monitoring attention may be especially important during social learning (Kumashiro et al. 2003; Rigamonti et al. 2005), social referencing (Russell et al. 1997) and cooperative problem solving (e.g. Hattori et al. 2005). Furthermore, in view of the cross-species differences in performance across tasks aimed at tapping into the same underlying ability to process gaze, greater efforts to increase the ecological validity of tasks will allow a better understanding of how species may benefit from exploiting the information conveyed by others' gaze.

14. Social Factors in Gaze Processing

One area that has been surprisingly neglected is the role of social status on gaze-monitoring, although it is known that social rank influences which group members primates attend to. Within a group, some individuals are monitored more than others; "attention structure" reflects social structure (Chance 1967). Subordinate individuals spend more time monitoring others than dominant group members do (Blois-Heulin and Girona 1999; McNelis and Boatright-Horowitz 1998; Watts 1998). Recent evidence suggests that status may also affect the tendency to co-orient with others: when viewing photographs of familiar group members, subordinate individuals showed reflexive co-orientation to all individuals, whereas dominants responded only to gaze shifts of other high-ranking individuals (Shepherd et al. 2006).

There are bound to be species differences in social influences on gaze-monitoring. Comparing a group of gelada baboons and mandrills, Emory (1976) found higher levels of social monitoring of and by the gelada male, but individual differences cannot be discounted. In another study, grey-cheeked mangabeys glanced (very brief fixation) at conspecifics more, whereas red-capped mangabeys usually fixated for longer (Blois Heulin 1999). It is conceivable that group members can extrapolate information regarding rank from the amount of attention an individual receives (Chance 1967), though this is a difficult hypothesis to test. The role of social status in mediating responses to gaze needs further exploration; we suggest that it should be considered as a factor in all gaze research.

15. Discussion and Conclusions

This selective overview of the literature on gaze-processing in primates has shown that different methodologies produce divergent findings, despite the fact that they all purportedly aim to explore the same basic abilities. Nonetheless, consistencies are emerging. Many species of nonhuman primates reliably and spontaneously demonstrate visual co-orientation in response to changes of both head and eye direction. In addition, apes also show an appreciation of opaque barriers when following gaze (Bräuer et al. 1995; Povinelli and Eddy 1996b; Tomasello et al. 1999), as may Old World monkeys, though the evidence is limited (Flombaum and Santos 2005). Nonhuman primates do not readily demonstrate gaze reading abilities within other contexts such as object-choice tasks. Nevertheless, they are able to learn contingencies between gaze information and outcomes, allowing them to solve the problems presented. More ecologically relevant approaches, such as food competition or social learning contexts, may be more suitable for revealing the role of gaze-monitoring in nonhuman primates. Although monkeys and apes may respond to eye gaze alone, like young human infants, they respond more readily to head orientation and other postural cues.

Whatever the experimental procedures used to study gaze-monitoring, the timing of intra-trial events is likely to be crucial. For example, Barth et al. (2005) tested 5 chimpanzees on an object choice task and reported better performance in a condition in which they approached an experimenter who was already giving a cue and made an immediate choice, than in a condition where they remained in the test area and waited for 10 sec while the cue was presented. It may be the case that in many studies using object-choice tasks; although the subjects may initially co-orient with a gaze cue, the delay between their co-orientation and object presentation may be too long. Also repetitive sessions may result in habituation to otherwise salient gaze behaviours (Povinelli and Eddy 1996a; Tomasello et al. 2001), and the presence of screens and objects that serve to obscure either the baiting events or the food itself may hinder performance (Vick and Anderson 2003).

Another methodological issue is the fact that in most gaze-monitoring studies the models are human. This means that factors such as social rank, sex, and quality of social relationship are usually neglected. Although they do respond to gaze shifts by humans, it seems likely that nonhuman primates are more attuned to gaze signals emitted by members of their own species. One exception to this may be those primates that have been reared in extensive contact with humans, the so-called enculturated primates (Tomasello and Call 1997). Those primates with more extensive interaction with humans respond more readily to experimenter given cues (Call and Tomasello 1994; Itakura and Tanaka 1998). It remains unclear whether enculturation results in fundamentally altered socio-cognitive abilities or simply a more intimate relationship with humans, which facilitates interactions both within and beyond the experimental context.

Because the gaze of others can reveal a great deal about their intentions and facilitate detection of features and events in both the social and non-social environments (Kummer 1967; Povinelli and Eddy 1996c; Tomasello et al. 1998), gaze processing can be studied as a primate adaptation, much as other socio-cognitive capacities (e.g. Schmidt and Cohn 2001). We need a better understanding of species typical gaze patterns and an appreciation of how gaze processing is mediated by other social factors (such as rank, age, sex) context (communication, competitive tasks, imitation, food sharing, and so on) and differences in socioecology between species, in order to test and generate hypotheses about the function of gaze.

References

Anderson JR, Doherty MJ (1997) Preschoolers' perception of other people's looking: photographs and drawings. Perception 26:333–343

Anderson JR, Mitchell RW (1999) Macaques but not lemurs co-orient visually with humans. Folia Primatol 70:17–22

Anderson JR, Sallaberry P, Barbier H (1995) Use of experimenter-given cues during object-choice tasks by capuchin monkeys. Anim Behav 49:201–208

Anderson JR, Montant M, Schmitt D (1996) Rhesus monkeys fail to use gaze direction as an experimenter-given cue in an object-choice task. Behav Processes 37:47–55

Anderson JR, Kuroshima H, Kuwahata H, Fujita K (2004) Do squirrel monkeys (*Saimiri sciureus*) and capuchin monkeys (*Cebus apella*) predict that looking leads to touching? Anim Cogn 7:185–192

Anderson JR, Kuroshima H, Hattori Y, Fujita K (2005) Attention to combined attention in New World monkeys (*Cebus apella, Saimiri sciureus*). J Comp Psychol 119:461–464

Anderson JR, Kuwahata H, Fujita K (2007) Gaze alternation during "pointing" by squirrel monkeys (*Saimiri sciureus*). Anim Cogn 10:267–271

Andrew RJ (1963) The origin and evolution of the calls and facial expressions of the primates. Behaviour 20:1–109

Anstis SM, Mayhew JW, Morley T (1969) The perception of where a face or television "portrait" is looking. Am J Psychol 82:474–489

Argyle M, Cook M (1976) Gaze and mutual gaze. Cambridge University Press, Cambridge

Argyle M (1988) Bodily communication, second edition. Routledge, London

Baron-Cohen S (1994) How to build a baby that can read minds: cognitive mechanisms in mindreading. Curr Psychol Cogn 13:513–552

Baron-Cohen S (1995). The eye director detector (EDD) and the shared attention mechanism (SAM): two cases for evolutionary psychology. In: Moore C, Dunham PJ (eds) Joint attention: its origins and role in development. Lawrence Erlbaum Associates, Hillsdale, NJ, pp 41–60

Baron-Cohen S, Wheelright S, Hill J, Raste Y, Plumb I (2001) The "Reading the Mind in the Eyes" test revised version: a study with normal adults, and adults with Asperger Syndrome or high-functioning Autism. J Child Psychol Psychiat 42:241–251

Barrett L, Henzi P, Dunbar, R (2003) Primate cognition: From "what now?" to "what if?" Trends Cogn Sci 7:494–497

Barth J, Reaux JE, Povinelli DJ (2005) Chimpanzees' (*Pan troglodytes*) use of gaze cues in object-choice tasks: different methods yield different results. Anim Cogn 8:84–92

Bates E, Camaioni L, Volterra V (1975) The acquisition of preformatives prior to speech. Merrill-Palmer Quart 21:205–226

Batki A, Baron-Cohen S, Wheelwright S, Connellan J, Ahluwalia J (2000) Is there an innate gaze module? Evidence from human neonates. Infant Behav Dev 23:223–229

Blois-Heulin C (1999) Variability in the social visual attention in the red-capped mangabey (*Cercocebus torquatus torquatus*) and the grey-cheeked mangabey (Cercocebus albigena albigena). Folia Primatol 70:264–268

Blois-Heulin C, Girona B (1999) Patterns of social visual attention in the red-capped mangabey (*Cercocebus torquatus torquatus*) in the context of food competition. Folia Primatol 70:180–184

Bräuer J, Call J, Tomasello M (2005) All great ape species follow gaze to distant locations and around barriers. J Comp Psychol 119:145–154

Brothers L (1996) Brain mechanisms of social cognition. J Psychopharmacol 10:2–8

Burghardt GM (1991). Cognitive ethology and critical anthropomorphism: a snake with two heads and hognose snakes that play dead. In: Ristau CA (ed) Cognitive ethology: the minds of other animals. Lawrence Erlbaum Associates, Hillsdale, NJ, pp 53–90

Butterworth G (1995). Origins of mind in perception and action. In: Moore C, Dunham PJ (eds) Joint attention: its origins and role in development. Lawrence Erlbaum Associates, Hillsdale, NJ, pp 29–40

Butterworth G, Jarret N (1991) What minds have in common is space: Spatial mechanisms serving joint visual attention in infancy. Br J Dev Psychol 9:55–72

Byrne RW, Whiten A (1988) Machiavellian intelligence: social expertise and the evolution of intellect in monkeys, apes, and humans. Clarendon Press, Oxford

Byrnit JT (2004) Nonenculturated orangutans' (*Pongo pygmaeus*) use of experimenter-given manual and facial cues in an object-choice task. J Comp Psychol 118:309–15

Call J, Tomasello M (1994) Production and comprehension of referential pointing by orangutans (*Pongo pymaeus*). J Comp Psychol 108:307–317

Call J, Hare BA, Tomasello M (1998) Chimpanzee gaze-following in an object-choice task. Anim Cogn 1:89–99

Call J, Agnetta B, Tomasello M (2000) Cues that chimpanzees do and do not use to find hidden objects. Anim Cogn 3:23–34

Campbell R, Heywood CA, Cowey A, Regard M, Landis T (1990) Sensitivity to eye gaze in prosopagnosic patients and monkeys with superior temporal sulcus ablation. Neuropsychologia 28:1123–1142

Chance MRA (1967) Attention structure as the basis of primate rank orders. Man 2:503–518

Cheney DL, Seyfarth RM (1991). Reading minds or reading behaviour? Tests for a theory of mind in monkeys. In: Whiten A (ed) Natural theories of mind: evolution, development and simulation of everyday mindreading. Blackwell, Cambridge, MA, pp 175–194

Corkum V, Moore C (1995). Development of joint visual attention in infants. In: Moore C, Dunham PJ (eds) Joint attention: its origins and role in development. Lawrence Erlbaum Associates, Hillsdale, NJ, pp 61–84

Coss RG (1978) Perceptual determinants of gaze aversion by the lesser mouse lemur (*Microcebus murinus*), the role of two facing eyes. Behaviour 64:248–67

Darwin C (1872) The expression of the emotions in man and animals. John Murray, London

Deaner RO, Platt ML (2003) Reflexive social attention in monkeys and humans. Curr Biol 13:1609–1613

Doherty-Sneddon G, Bruce V, Bonner L, Longbotham S, Doyle C (2002) Development of gaze aversion as disengagement from visual information. Dev Psychol 38:438–445

Eacott MJ, Heywood CA, Gross CG, Cowey C (1993) Visual discrimination impairments following lesions of the superior temporal sulcus are not specific for facial stimuli. Neuropsychologia 31:609–619

Ekman P, Rosenberg E (2005) What the face reveals (2nd ed.). Oxford University Press, New York

Emery N (2000) The eyes have it: the neuroethology, function and evolution of social gaze. Neurosci Behav Rev 24:581–604

Emery NJ, Lorincz EN, Perret DI, Oram MW, Baker CI (1997) Gaze-following and joint attention in rhesus monkeys (*Macaca mulatta*). J Comp Psychol 111:286–293

Emory GM (1976) Aspects of attention, orientation, and status hierarchy in mandrills (*Mandrillus sphinx*) and gelada baboons (Theropithecus gelada). Behaviour 59:70–87

Exline RV, Yellin AM (1969). Eye contact as a sign between man and monkey. *Symposium on Non-verbal Communication*, Nineteenth International Congress of Psychology, London

Fagot J, Deruelle C (2001) Perception of pictorial gaze by baboons (*Papio papio*). J Exp Psychol: Anim Behav Process 28:298–308

Farroni T, Csibra G, Simion F, Johnson MH (2002) Eye contact detection in humans from birth. Proc Nat Acad Sci USA 99:9602–9605

Ferrari PF, Kohler E, Fogassi L, Gallese V (2000) The ability to follow eye gaze and its emergence during development in macaque monkeys. Proc Natl Acad Sci 97:13997–14002

Flombaum JI, Santos LR (2005) Rhesus monkeys attribute perceptions to others. Curr Biol 15:447–452

Frank M, Ekman P, Priesen W (1993) Behavioral markers and recognizability of the smile of enjoyment. J Pers Soc Psychol 64:83–93

Friesen CK, Kingstone A (1998) The eyes have it! Reflexive orienting is triggered by nonpredictive gaze. Psychonom Bull Rev 5:490–495

Gallup GG Jr (1972) Some chickens I have intimidated. Psychol Today 6:62–64
Gibson JJ, Pick AD (1963) Perception of another person's looking behaviour. Am J Psychol 76:386–394
Glenberg M, Schroeder JL, Robertson DA (1998) Averting the gaze disengages the environment and facilitates remembering. Mem Cogn 26:651–658
Gómez JC (1996). Ostensive behaviouor in the great apes. In: Russon AE, Bard KA, Parker ST (eds) Reaching into thought: the minds of the great apes. Cambridge University Press, Cambridge, pp 131–151
Gordon RM (1998) The prior question: do *human* primates have a theory of mind? Behav Brain Sci 21:120–121
Gothard KM, Erickson CA, Amaral DG (2004) How do rhesus monkeys (*Macaca mulatta*) scan faces in a visual paired comparison task? Anim Cogn 7:25–33
von Grünau M, Anston C (1995) The detection of gaze direction: a stare in the crowd effect. Perception 24:1297–1313
Guo K, Robertson RG, Mahmoodie S, Tadmor Y, Young MP (2003) How do monkeys view faces? A study of eye movements. Exp Brain Res 150:363–374
Gygax L (1995) Hiding behaviour of longtailed macaques (*Macaca fascicularis*): I. Theoretical background and data on mating. Ethology 101:10–24
Gygax L (2000) Hiding behaviour of long-tailed macaques (*Macaca fascicularis*): II. Use of hiding places during aggressive interactions. Ethology 106:441–451
Hampton RR (1994) Sensitivity to information specifying the line of gaze of humans in sparrows (*Passer domesticus*). Behaviour 130:41–51
Hare B (2001) Can competitive paradigms increase the validity of experiments of primate social cognition? Anim Cogn 4:269–280
Hare B, Call J, Agnetta B, Tomasello M (2000) Chimpanzees know what conspecifics do and do not see. Anim Behav 59:771–785
Hare B, Call J, Tomasello M (2001) Do chimpanzees know what conspecifics know? Anim Behav 61:139–151
Hare B, Addessi E, Call J, Tomasello M, Visalberghi E (2003) Do capuchin monkeys, *Cebus apella*, know what conspecifics do and do not see? Anim Behav 65:131–142
Hare B, Tomasello M (2004) Chimpanzees are more skilful in competitive than in cooperative cognitive tasks. Anim Behav 68:571–581
Hattori Y, Kuroshima H, Fujita K (2005) Cooperative problem solving by tufted capuchin monkeys (*Cebus apella*): Spontaneous division of labor, communication, and reciprocal altruism. J Comp Psychol 119:335–342
Heyes CM (1998) Theory of mind in nonhuman primates. Behav Brain Sci 21:101–148
Hietanen JK (2002) Social attention orienting integrates visual information from head and body orientation. Psychol Res 66:174–179
Hood BM, Willen JD, Driver J (1998) Adult's eyes trigger shifts of visual attention in human infants. Psychol Sci 9:131–134
van Hooff JARAM (1967) The facial displays of catarrhine monkeys and apes. In: Morris D (ed) Primate ethology. Weidenfeld and Nicolson, London, pp 7–68
Humphrey N (1976). The social function of intellect. In: Bateson PPG, Hinde RA (eds) Growing points in ethology. Cambridge University Press, Cambridge, pp 303–317
Inoue Y, Inoue E, Itakura S (2004) Use of experimenter-given directional cues by a young white-handed gibbon (*Hylobates lar*). Jpn Psychol Res 46:262–267
Itakura S (1996) An exploratory study of gaze-monitoring in nonhuman primates. Jpn Psychol Res 38:174–180

Itakura S, Anderson JR (1996) Learning to use experimenter-given cues during an object-choice task by a capuchin monkey. Curr Psychol Cogn 15:103–112

Itakura S, Tanaka M (1998) Use of experimenter-given cues during object-choice tasks by chimpanzees (*Pan troglodytes*), an orangutan (*Pongo pygmaeus*) and human infants (*Homo sapiens*). J Comp Psychol 112:119–126

Itakura S, Agnetta B, Hare B, Tomasello M (1999) Chimpanzee use of human and conspecific social cues to locate hidden food. Dev Sci 2:448–456

Kalin NH, Shelton SE (1989) Defensive behaviors in infant rhesus monkeys: Environmental cues and neurochemical regulation. Science 243:1718–1721

Kalin NH, Shelton SE, Takahashi LK (1991) Defensive behaviors in infant rhesus monkeys: ontogeny and context-dependent selective expression. Child Dev 62:1175–1183

Kampe KKW, Frith CD, Dolan RJ, Frith U (2001) Reward value of attractiveness and gaze. Nature 413:589

Kaplan G, Rogers LJ (2002) Patterns of eye gazing in orangutans. Int J Primatol 23:501–526

Keating CF, Keating EG (1982) Visual scan patterns of rhesus monkeys viewing faces. Perception 11:211–219

Keating CF, Keating EG (1993) Monkeys and mug shots: cues used by rhesus monkeys (*Macaca mulatta*) to recognize a human face. J Comp Psychol 107:131–139

Keltner D (1995) Signs of appeasement: Evidence for the distinct displays of embarrassment, amusement, and shame. J Pers Soc Psychol 68:441–454

Kingstone A, Friesen CK, Gazzaniga MS (2000) Reflexive joint attention depends on lateralized cortical connections. Psychol Sci 11:1–8

Kingstone A, Tipper C, Ristic J, Nganb E (2004) The eyes have it!: an fMRI investigation. Brain Cogn 55:269–271

Kleinke C (1986) Gaze and eye contact: a research review. Psychol Bull 100:78–100

Kobayashi H, Kohshima S (1997) Unique morphology of the human eye. Nature 387:767–768

Kobayashi H, Kohshima S (2001) Unique morphology of the human eye and its adaptive meaning: comparative studies of external morphology of the primate eye. J Hum Evol 40:419–435

Kumashiro M, Ishibashi H, Itakura S, Iriki A (2002) Bidirectional communication between a Japanese monkey and a human through eye gaze and pointing. Curr Psychol Cogn 21:3–32

Kumashiro M, Ishibashi H, Uchiyama Y, Itakura S, Muratac A, Iriki A (2003) Natural imitation induced by joint attention in Japanese monkeys. Int J Psychophysiol 50: 81–99

Kummer H (1967). Tripartite relationships in Hamadryas baboons. In: Altmann SA (ed) Social communication among primates. University of Chicago Press, Chicago, IL, pp 56–63

Kummer H, Anzenberger G, Hemelrijk CK (1996) Hiding and perspective taking in long-tailed macaques. J Comp Psychol 110:97–102

Kuroshima H, Fujita K, Fuyuki A, Masuda T (2002) Understanding of the relationship between seeing and knowing by tufted capuchin monkeys (*Cebus apella*). Anim Cogn 5:41–48

Kyes RC, Candland DK (1987) Baboon (*Papio hamadryas*) visual preferences for regions of the face. J Comp Psychol 101:345–348

Langton SRH, Watt RJ, Bruce V (2000) Do the eyes have it? Cues to the direction of social attention. Trends Cogn Sci 4:50–59

Leavens DA, Hopkins WD, Bard KA (1996) Indexical and referential pointing in chimpanzees (*Pan troglodytes*). J Comp Psychol 110:346–353

Liebal K, Pika S, Call J, Tomasello M (2004) To move or not to move: how apes alter the attentional state of others. Interact Stud 5:199–219

Lorincz EN, Baker CI, Perret DI (1999) Visual cues for attention following in rhesus monkeys. Curr Psychol Cogn 18:973–1003

Mason MF, Tatkow EP, Macrae CN (2005) The look of love: gaze shifts and person perception. Psychol Sci 16:236–239

McNelis NL, Boatright-Horowitz SL (1998) Social monitoring in a primate group: the relationship between visual attention and hierarchical ranks. Anim Cogn 1:65–69

Mendelson MJ, Haith MM, Golman-Rakic PS (1982) Face scanning and responsiveness to social cues in infant rhesus monkeys. Dev Psychol 18:222–228

Menzel EW (1974). A group of young chimpanzees in a one acre field: leadership and communication. In: Schrier AM, Stollnitz F (eds) Behavior of nonhuman primates. Academic Press, New York, pp 83–153

Montgomery DE, Bach LM, Moran C (1998) Children's use of looking behaviour as a cue to detect another's goal. Child Dev 69:692–705

Moore C (1999). Gaze-following and the control of attention. In: Rochat P (ed) Early social cognition: understanding others in the first months of life. Lawrence Earlbaum Associates, London, pp 241–256

Morris D (1967) The naked ape. Jonathan Cape, London

Myowa-Yamakoshi M, Tomonaga M (2001) Perceiving eye gaze in an infant gibbon (*Hylobates agilis*). Psychologia 44:24–30

Myowa-Yamakoshi M, Tomonaga M, Tanaka M, Matsuzawa T (2003) Preference for human direct gaze in infant chimpanzees (*Pan troglodytes*). Cognition 89:B53–B64

Nahm FK, Perret A, Amaral DG, Albright TD (1997) How do monkeys look at faces? J Cogn Neurosci 9:611–623

Neiworth JJ, Burman MA, Basile BM, Lickteig MT (2002) Use of experimenter given cues in visual coorienting and in an object choice task by a New World monkey species, cotton top tamarins (*Saguinis oedipus*). J Comp Psychol 112:119–126

Okamoto S, Tomonaga M, Ishii K, Kawai N, Tanaka M, Matsuzawa T (2002) An infant chimpanzee (*Pan troglodytes*) follows human gaze. Anim Cogn 5:107–114

Parr LA, de Waal FBM (1999) Visual kin recognition in chimpanzees. Nature 399: 647–648

Parr LA, Winslow JT, Hopkins WD, de Waal FBM (2000) Recognising facial cues: individual discrimintaion by chimpanzees (*Pan troglodytes*) and rhesus monkeys (*Macaca mulatta*). J Comp Psychol 114:47–60

Pascalis O, Petit O, Kim JH, Campbell R (1999) Picture perception in primates: the case of face perception. In: Fagot J (ed) Picture perception in animals. Psychology Press/ Taylor & Francis, Hove, pp 263–294

Peignot P, Anderson JR (1999) Use of experimenter given manual and facial cues by gorillas (*Gorilla gorilla*) in an object-choice task. J Comp Psychol 113:253–260

Perrett DI, Emery NJ (1994) Understanding the intentions of others from visual signals: neurophysiological evidence. Curr Psychol Cogn 13:683–694

Perrett DI, Mistlin AJ (1990). Perception of facial characteristics by monkeys. In: Stebbins WC, Berkley MA (eds) Comparative perception, Volume II: Complex signals. John Wiley & Sons, New York, pp 187–213

Povinelli DJ, Eddy TJ (1996a) What young chimpanzees know about seeing. University of Chicago Press, Chicago, IL

Povinelli DJ, Eddy TJ (1996b) Factors influencing young *chimpanzees*' (*Pan troglodytes*) recognition of attention. J Comp Psychol 110:336–345

Povinelli DJ, Eddy TJ (1996c) Chimpanzees: joint visual attention. Psychol Sci 7: 129–135

Povinelli DJ, Eddy TJ (1997) Specificity of gaze-following in young chimpanzees. Br J Dev Psychol 15:213–222

Povinelli DJ, Giambrone S (2000) Escaping the argument by analogy. In: Povinelli DJ (ed) Folk physics for apes. Oxford University Press, Oxford, pp 9–72

Povinelli DJ, Nelson KE, Boysen ST (1990) Inferences about knowing and guessing by chimpanzees (*Pan troglodytes*). J Comp Psychol 104:203–210

Povinelli DJ, Biershwale DT, Cech CG (1999) Comprehension of seeing as a referential act in young children, but not juvenile chimpanzees. Br J Dev Psychol 17:37–60

Premack D (1988) "Does the chimpanzee have a theory of mind" revisited. In: Byrne RW, Whiten A (eds) Machiavellian intelligence: social expertise and the evolution of intellect in monkeys, apes, and humans. Clarendon Press, Oxford, pp 160–179

Preuschoft S (2000) Primate faces and facial expressions. Soc Res 67:245–271

Ricclardelli P, Baylis G, Driver J (2000) The positive and negativite of human expertise in gaze perception. Cognition 77:B1–B14

Rigamonti MM, Custance DM, Previde EP, Spiezio C (2005) Testing for localized stimulus enhancement and object movement reenactment in pig-tailed macaques (*Macaca nemestrina*) and young children (*Homo sapiens*). J Comp Psychol 119:257–272

Ristau CA (1998). Cognitive ethology: the minds of children and animals. In: Cummins DD, Allen C (eds) The evolution of mind. Oxford University Press, New York, pp 127–161

Ristic J, Friesen CK, Kingstone A (2002) Are eyes special? It depends on how you look at it. Psychon Bull Rev 9:507–513

Roberts SC, Havlicek J, Flegr J, Hruskova M, Little AC, Jones BC, Perrett DI, Petrie M (2004) Female facial attractiveness increases during the fertile phase of the menstrual cycle. Proc R Soc Lond B 271:270–272

Russell CL, Bard KA, Adamson LB (1997) Social referencing by young chimpanzees (*Pan troglodytes*). J Comp Psychol 111:185–193

Russell JL, Braccini S, Buehler N, Kachin MJ, Schapiro SJ, Hopkins WD (2005) Chimpanzee (*Pan troglodytes*) intentional communication is not contingent upon food. Anim Cogn 8:263–272

Santos LR, Hauser MD (1999) How monkeys see the eyes: cotton top tamarins' reaction to changes in visual attention and action. Anim Cogn 2:131–139

Sato M, Nakamura K (2001) Detection of direct gaze in rhesus monkeys (*Macaca mulatta*). J Comp Psychol 115:115–121

Scaife M, Bruner J (1975) The capacity for joint attention in the infant. Nature 253: 265–266

Scerif G, Gomez JC, Byrne RW (2004) What do Diana monkeys (*Cercopithecus diana diana*) know about conspecifics' focus of attention? Anim Behav 68:1239–1247

Schmidt KL, Cohn JF (2001) Human facial expressions as adaptations: evolutionary questions in facial expression research. Yearb Phys Anthropol 44:3–24

Senju A, Tojo Y, Dairoku H, Hasegawa T (2004) Reflexive orienting in response to eye gaze and an arrow in children with and without autism. J Child Psychol Psychiat 45:445–458

Setchell JM, Dixson AF (2001) Changes in secondary sexual adornments of male mandrills (*Mandrillus sphinx*) are associated with gain and loss of alpha status. Horm Behav 9:177–184

Seyama J, Nagayama RS (2002) Perceived eye size is larger in happy faces than in surprised faces. Perception 31:1153–1155

Shepherd SV, Deaner RO, Platt ML (2006) Social status gates social attention in monkeys. Curr Biol 16:R119–R120

Swettenham J, Condie S, Campbell R, Milne E, Coleman M (2003) Does the perception of moving eyes trigger reflexive visual orienting in autism? Philos Trans R Soc Lond B 358:325–334

Theall LA, Povinelli DJ (1999) Do chimpanzees tailor their gestural signals to fit the attentional states of others? Anim Cogn 2:207–214

Thomsen CE (1974) Eye contact by non-human primates towards a human observer. Anim Behav 22:144–149

Tomasello M (1995). Joint attention as social cognition. In: Moore C, Dunham PJ (eds) Joint attention: its origins and role in development. Lawrence Erlbaum Associates, Hillsdale, NJ, pp 103–130

Tomasello M, Call J (1997) Primate cognition. Oxford University Press, Oxford

Tomasello M, Call J, Nagell K, Olguin K, Carpenter M (1994) The learning and use of gestural signals by young chimpanzees. Primates 35:137–154

Tomasello M, Call J, Hare B (1998) Five primate species follow the visual gaze of conspecifics. Anim Behav 55:1063–1069

Tomasello M, Hare B, Agnetta B (1999) Chimpanzees (*Pan troglodytes*) follow gaze direction geometrically. Anim Behav 58:769–777

Tomasello M, Hare B, Fogleman T (2001) The ontogeny of gaze-following in chimpanzees (*Pan troglodytes*) and rhesus macaques (*Macaca mulatta*). Anim Behav 61:335–343

Vick S.-J., Anderson JR (2000) Learning and limits of use of eye gaze by capuchin monkeys (*Cebus apella*) in an object-choice task. J Comp Psychol 114:200–207

Vick S.-J., Anderson JR (2003) The use of visual orientation cues in a competitive task by olive baboons (*Papio anubis*). J Comp Psychol 117:209–216

Vick S.-J., Bovet D, Anderson JR (2001) Discrimination of gaze cues by olive baboons (*Papio anubis*). Anim Cogn 4:1–10

Walden TA, Ogan TA (1988) The development of social referencing. Child Dev 59: 1230–1240

Watts DP (1998) A preliminary study of selective visual attention in female mountain gorillas (*Gorilla gorilla beringei*). Primates 39:71–78

Werner H, Kaplan B (1963) Symbol formation. Wiley, New York

Woodruff G, Premack D (1979) Intentional communication in the chimpanzee: the development of deception. Cognition 7:333–362

4
How to Build a Scrub-Jay that Reads Minds

Nathan J. Emery[1] and Nicola S. Clayton[2]

1. Introduction

Although the search for human-like theory of mind (ToM) in non-human animals has continued unabated for almost 30 years, we have made very little progress in determining the psychological processes involved in non-human social cognition compared to the significant progress made in human infants. The underlying assumption that forms the basis for this research, is that ToM did not appear de novo in the evolutionary record of humans. Therefore, other animals, and most notably monkeys and apes, should at least demonstrate some of the precursors of socio-cognitive processing demonstrated by *Homo sapiens*. For example, there is good evidence that many animals follow another's line of sight to external objects (e.g. monkeys; apes; dolphins; domestic dogs; goats; seals; ravens; Grey parrots, see Emery 2000 for review). However, this ability need not be explained in mentalistic terms, as in many cases it may be purely reflexive (Povinelli and Eddy 1996). Indeed, this simpler explanation seems likely based on the generality of the behaviour, and the number of species in which it has been demonstrated.

The presence or absence of ToM in non-human animals becomes more controversial when we examine more sophisticated forms of social cognition, such as visual perspective-taking, knowledge attribution and false-belief (Emery 2005). Many chapters in this volume will attest to the sophisticated mind-reading skills of a wide variety of species, however, the plain and simple truth is that no non-human animal will ever demonstrate human ToM. As such, we should change our experimental focus onto what constitutes a species-specific social cognition. This is the aim of this chapter.

2. Why Comparative Social Cognition may be Failing?

We suggest that there are three reasons why comparative social cognition may be failing to achieve its goals. First, there appears to be little appreciation of the 3Es (ecology, ethology and evolution), both in the design of experiments and in

[1]School of Biological and Chemical Sciences, Queen Mary University of London, Mile End Road, London E1 4NS, UK
[2]Department of Experimental Psychology, University of Cambridge, Downing Street, Cambridge CB2 3EB, UK

thinking about what social cognition might be used for in the natural behaviour of an animal. By ecology, we mean environmental constraints on cognition, such as diet, habitat as well as social/mating system. Perhaps the best example is the case of food-caching corvids in which there are differences between species in performance on a variety of spatial memory tasks, which reflect differences in the ecology of the species (Balda et al. 1996). Clark's nutcrackers live at high altitudes in which food is scarce during the long, cold winter, they tend to eat and cache a very restricted diet (pine seeds), and they cache up to 30,000 pine seeds each year. By contrast, western scrub-jays live in a much more temperate climate at a much lower elevation, with a more varied diet, and which cache a lot fewer pine seeds (approx. 6,000 per year).

In laboratory tests for spatial memory, including cache retrieval, the Clark's nutcrackers tend to outperform the western scrub-jays, demonstrating greater accuracy for remembering cache location after long periods, or performing well on other spatial memory tasks not related to caching. However, Clark's nutcrackers and western scrub-jays not only differ on the amount of seeds cached and the habitat in which they live, but also in sociality. Clark's nutcrackers are territorial, and solitary (except during the breeding season), whereas western scrub-jays are semi-territorial and often form small flocks outside of the breeding season (Balda et al. 1996). Both Clark's nutcrackers and western scrub-jays are relatively asocial when compared to the closely-related pinyon jay, which routinely gather in flocks of 80 individuals, forming linear dominance hierarchies (Balda 2002).

Based on this ecological information, we would predict that in tests of sociality, pinyon jays should outperform both Clark's nutcrackers and western scrub-jays. Pinyon jays appear to use transitive inference to represent conspecifics' relative dominance status (Paz-y-Mino et al. 2004). Furthermore, they also seem to learn non-social versions of a transitive inference problem faster than western scrub-jays (Bond et al. 2003), however the scrub-jays eventually learn the task, and in the same way as the pinyon jays.

Furthermore, in a test of social cognition; observational spatial memory (which allows a cache pilferer to accurately locate another's caches); Clark's nutcrackers successfully locate another's caches even after a 1-day interval. They are less successful at a 2-day interval, compared to pinyon jays which are successful at this longer delay (Bednekoff and Balda 1996a,b). Western scrub-jays have only been tested at relatively short delays (3 hr), but are almost as accurate as the storers in locating their caches (Clayton et al. 2001). Data of this sort suggest that ecological information is necessary, but not sufficient for designing and interpreting biologically valid tests of comparative social cognition.

An appreciation of evolution is also important when considering species differences in cognitive ability. Let us return to the comparison between Clark's nutcrackers and pinyon jays. A phylogenetic reconstruction of the evolution of caching by corvids suggests that intense caching evolved independently in these two different lineages of corvids, and consequently there is no a priori reason to assume that the two species have developed the same solution to the problem of having to remember the location of so many food caches (de Kort and Clayton

2006). Consequently, the two species might have different problems when tested on tasks that rely on spatial memory, and closely related species might show more similar levels of performance than more distantly related ones.

To illustrate the point, we shall consider two comparative studies of spatial memory in corvids. In the first, Olson and colleagues compared pinyon jays and Clark's nutcrackers with the western scrub-jay and the Mexican jay, both of whom are moderate cachers, for their performance on an operant spatial delayed non-matching to sample (DNMTS) task using a touch screen (Olson et al. 1995), the pinyon jays were no better at this task than the other two jay species. The pinyon jay's performance was more similar to those of its close relatives, the scrub-jays and Mexican jays, despite being a specialised food-storer like the Clark's nutcracker.

In a second study, Gould-Beierle (2000) compared the performance of Clark's nutcrackers, pinyon jays, western scrub-jays and jackdaws on a radial maze. The Clark's nutcracker's performance was no better that that of the non-caching jackdaw, while both were outperformed by the pinyon jay and scrub-jay. As in the previous example, de Kort and Clayton (2006) point out that explanatory power comes from considerations of evolutionary relatedness: the nutcrackers are more closely related to jackdaws than they are to the two species of New world jay, despite the fact that Clark's nutcrackers are intense storers and known for their remarkable spatial performance in a variety of other spatial tasks.

The third E is ethology, and clearly understanding an animal's ethology or their natural behavioural repertoire is also important. In the case of social cognition, this means recognising examples of natural behaviour that may require mind-reading or in which social cognition provides an adaptive advantage over not having socio-cognitive abilities. Examples that have formed the basis for successful research programs are food competition in chimpanzees, caching and pilfering in corvids, and dog-human communication. We will focus our discussion to cache protection in corvids. Other potential examples, which have yet to be exploited by researchers, are cooperation in the formation of coalitions and alliances, and redirected aggression.

A second difficulty with comparative social cognition research has been the bias towards comparisons with humans, and consequently a focus on folk or naïve psychology (theory ToM) rather than alternative forms of social cognition, such as simulation ToM. This approach may have set up a "straw man" by which all non-linguistic creatures (non-humans and preverbal infants) will fail, as this approach is dependent on the development of language (Smith 1996). We will argue later in the chapter that the alternative simulation ToM approach may be better to test for social cognition in pre-verbal or non-verbal creatures.

Finally, until very recently comparative social cognition was not particularly comparative. Studies tended to be restricted to a small number of species, largely monkeys and apes. This has changed a little in the last few years, including work on domestic dogs (Hare and Tomasello 2005), foxes (Hare et al. 2005), domestic goats (Kaminski et al. 2005), dolphins (Tschudin 2006) and various species of birds (see below). However, if we are to learn what were the evolutionary precursors to the socio-cognitive skills of humans, whether social cognition takes different forms in different taxa, and how it develops with and across species, we need to expand the range of species being studied to closely related (comparative), and distantly related (phylogenetic) species, as well as individuals at different stages of development (ontogenetic; Tomonaga 2006).

The last 10 years has seen an increase in the use of the ecological approach in the design of animal cognition experiments (Balda et al. 1996). This is especially important for those studies on species distantly related to humans, as it is not clear whether such experiments are directly comparable in species with different perceptual worlds (Bitterman 1975). The ecological approach provides an opportunity for the animal under investigation to demonstrate the "best" of its abilities, rather than an anthropocentric bias inherent in studies designed for human and non-human primates. This approach, by increasing ecological validity, also matches the animal's behaviour in the wild more closely than any study in an information poor environment using arbitrary stimuli (Kamil and Maudlin 1988; Shettleworth 1998).

3. Avian Social Cognition

Although many birds form large, complex social groups, and form close relationships with other group members that resemble those found in some mammals, notably primates, elephants and cetaceans (Emery et al. 2007), the cognitive mechanisms by which birds may process social information (avian social cognition) has only recently become a viable topic for study. This is largely because of recent changes in our perception of birds as cognitive creatures.

The avian brain has, for over 100 years, been seen as a much simpler structure than the mammalian brain. The vast majority of the mammalian brain is cortical, neural tissue formed into six-layers of cells with both vertical and horizontal connections. The neocortex and particularly the prefrontal cortex, is involved in cognitive processing; thinking, memory, planning and social interaction. The neocortex was adapted from the pallium, whereas the equivalent area of the avian forebrain was traditionally seen as being adapted from the striatum (basal ganglia; Striedter 2005). The basal ganglia is involved in species-typical behaviour (sex, parenting, feeding, etc.) which is unlikely to be based on cognitive processing. Therefore the potential for intelligence in birds was seen as being constrained by their brains. However, recent data on the evolution, structure, neurochemical composition and function of the avian forebrain has demanded a re-evaluation of the cognitive skills of birds. Indeed, the majority of the avian forebrain has been renamed as pallial rather than striatal based on these studies (Avian Brain Nomenclature Consortium 2005; Emery and Clayton 2005; Reiner et al. 2004), suggesting that birds do have the brains to support the complex forms of cognition so-far only described for large-brained mammals.

Not all birds were created equal. Some families of birds, such as *Corvidae* (crows and jays) and *Psittacidae* (parrots) share many biological, ecological, neurobiological and behavioural traits with primates (Emery 2006); as such they may provide an example of convergent evolution (e.g. evolving analogous solutions to a similar problem), whereas others do not. We have suggested that the cognitive abilities of corvids and apes may also represent a case for convergent evolution, but with divergent evolution of neural systems (Emery and Clayton 2004a,b). As such, this provides us with a theoretical standpoint to investigate more complex forms of cognitive processing than would be possible if we had to base our investigations on out-dated models of comparative psychology that do not consider the 3Es and assume that all animals are intellectually equal (MacPhail 1987).

The study of social cognition in birds has had a relatively short history, largely because of the species of birds tested (pigeons, quail or chickens), the negative bias against complex cognition in birds because of their presumed small brain size and suggested limited flexibility in behaviour and because of a primatocentric bias. Although some of the best evidence for social learning comes from birds (Lefebvre and Bouchard 2003; Zentall 2004), the fact that birds are so distantly related to humans has made the design of suitable experiments a difficult stumbling block to progress. Also, contrary to the views of many, some species of birds demonstrate complex patterns of behaviour that are deployed flexibly depending on the context (e.g. innovation and tool use; Lefebvre et al. 1997, 2002).

Early studies found that birds perceived eyes and eye-like stimuli as aversive (Jones 1980; Scaife 1976a,b) supposedly representing the frontal orientation of a predator. Although chickens react strongly to the presence of two eyes and less so to one eye (maybe representing a predator looking away), frontal head orientation appears to be a sufficient cue to elicit strong aversive reactions in sparrows (Hampton 1994).

Studies on more complex aspects of social cognition, which have been tested on primates and other social mammals, did not occur until very recently (largely because of the reasons described above). For example, ravens can follow the gaze direction of a human experimenter behind a barrier (Bugnyar *et al.* 2004). Bee-eaters react to the visual perspective of a human "predator" approaching their nest from different directions, although this is likely based on computing the orientation of the human's face in relation to the nest rather than any understanding of what humans can or cannot see (Watve et al. 2002). African grey parrots appear to use a human's attention to learn about the names of objects (Pepperberg and McLaughlin 1996). In corvids at least, a sudden increase in interest in the socio-cognitive skills of birds can be attributed to a shift in using the caching paradigm to examine cache protection strategies as opposed to spatial memory. These experiments will form the basis for the rest of this chapter.

4. Cache Protection Strategies as Social Cognition

4.1. Biology of Cache Protection

Food-storing corvids, such as jays and ravens, cache food for future consumption and rely on memory to recover their caches at a later date. Food caching and recovery are activities which occur within a social context, not least because caches are susceptible to pilfering by other individuals, both other food-storers and also non-storing heterospecifics such as jackdaws (Dally et al. 2006b; Vander Wall 1990). For pilferers, the ability to locate caches made by others quickly and efficiently may be the important difference between successful pilfering and potential aggression from the storer. So an obvious advantage of observational learning of food cache location is that it allows birds to efficiently pilfer caches when others have left the scene, thereby eliminating both the costs of caching and of fighting (Clayton and Emery 2004). A number of corvids observe the caching behaviour of their fellow conspecifics, and show excellent spatial memory in locating another bird's food caches (e.g., Bednekoff and Balda 1996a,b; Bugnyar and Kotrschal 2002; Clayton et al. 2001; Heinrich and Pepper 1998).

The social context of caching behaviour may be viewed as a "cognitive arms race" between storers and pilferers, in which the storers use counter strategies to minimise the risk of having their caches pilfered (Bugnyar and Kotrschal 2002; Dally et al. 2006b; Emery et al. 2004). In this arms race, however, an individual bird can play both roles. Indeed, in highly colonial food-storers such as rooks, any member of a social group may play the role of both storer of its own caches and stealer of other individuals' caches depending on the circumstances. Field observations suggest that the storers engage in a number of cache protection strategies such as waiting until the would-be pilferers are distracted or cannot see them before they resume caching, or by making "false" caches that either contain a inedible item such as a stone or nothing at all (e.g. rooks, Kallander 1978; ravens, Heinrich 1999; Heinrich and Pepper 1998; Bugnyar and Kotrschal 2002). And, some corvids return alone to caches they had hidden in the presence of conspecifics, and readily re-cache them in new places unbeknown to the potential thief (e.g. jays, Emery and Clayton 2001, ravens, Heinrich 1999).

While field observations are of enormous value in documenting the natural behaviour of these birds, an experimental approach is crucial for understanding the mechanisms underlying these behaviours and determining the effects of experience, particularly in relation to "theory of mind" (Clayton et al. 2007). Consider the observation of birds moving the food they had hidden in the presence of other individuals, and re-caching the food items in new places when those observers were no longer present. In the wild, one might explain the presence or absence of another bird as purely coincidental to the caching and re-caching events. To test whether it is the presence of an observer at caching, and absence of one at recovery, that elicts the food-cacher's re-caching behaviour, Emery and Clayton (2001) allowed hand-raised western scrub-jays to cache either in private or while a conspecific was watching and then recover their caches in private. Individuals subquently re-cached food in new cache sites, but only when they had been observed during caching (Fig. 1). We argue that because the two conditions were identical at the time of recovery (in private), the birds had to remember whether or not they had been watched during the caching condition in order to know when to re-cache during recovery, and whether to re-cache in new sites. By focussing on the counter strategies of the storer when previously observed by

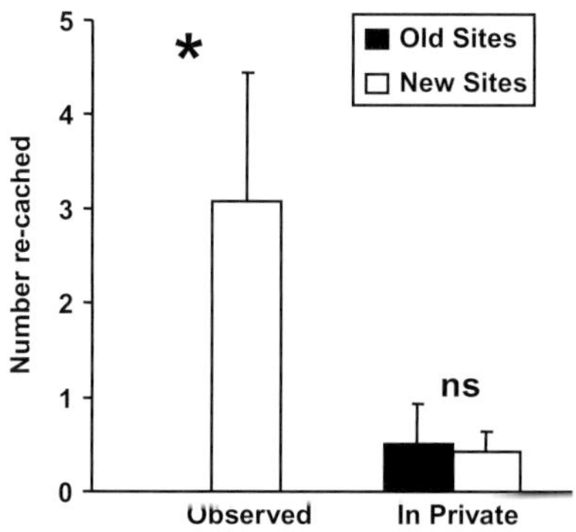

FIG. 1. Number of food items re-cached by storers during recovery in private; either when they had been observed during caching or had cached in private. Items were re-cached into either old sites (in the same tray) or new sites (different tray not made previously available). (Adapted from Emery and Clayton 2001, $*p < 0.05$)

a potential thief, this experiment raises the intriguing possibility that this re-caching behaviour is based on mental attribution.

Bugnyar and Kotrschal (2002) have also suggested that the storing and stealing tactics used by some food-caching corvids may involve complex social cognition. These authors suggest that ravens engage in tactical deception through an attempt to manipulate another's attention, either to prevent opponents from gaining opportunities to steal, or to gain opportunities for learning socially from their opponents. Raven cachers try to increase the distance between themselves and potential thieves, and often halted caching or re-cached based on the behaviour of the observers; whereas raiders also increased their distance from cachers. Bugnyar and Kotrschal (2002) argued that raven cachers appear to appreciate the visual perspective of potential raiders because they cache behind large objects (e.g. tree stumps and rocks) and position themselves in relation to the object and the observer so preventing the observer from ⟨seeing⟩ them cache. It is not clear, however whether the storers appreciate the concept ⟨see⟩ in others or whether they are merely responding to the fact that they cannot ⟨see⟩ the observer. This will be discussed further in the next section.

4.2. Visual Perspective Taking

A bird which caches in front of a conspecific may need to appreciate that individual's visual perspective to determine whether the potential thief will be able to use observational spatial memory to locate those caches at a later time. It would be inefficient to utilise protection behaviours with respect to caches which are not in danger, such as those that have not been seen. At the time of caching, a storer may assess the amount of visual information available to a pilferer and may either reduce that information *during* caching or move caches during recovery which have been *seen* during caching. We have investigated such cache protection strategies in three studies.

In the first experiment, scrub-jays were allowed to cache either in front of a conspecific or in private where the view of the conspecific was completely occluded. In both conditions the storer was presented with a choice of where to cache; either in a caching tray located in a well-lit part of the cage or in a caching tray located in a darkened part of the cage (Fig. 2a; Dally et al. 2004). The suggestion being that it will be more difficult for an observing scrub-jay to see where caches have been placed in the darkened tray compared to caches placed in the well-lit tray. Even though caches are made in restricted localisations (i.e. caching trays with only 8 potential cache sites), pilfering is a dangerous business, especially when the storer is more dominant to you. Therefore, pilferers need to steal caches quickly and accurately.

Although the scrub-jays' favourite food is wax worms, we provided them with pine nuts in this experiment, so that degradation of the worms under the heat of the sun lamp used to provide light to one side of the cage could not explain their behaviour. When the jays cached in private they cached equal amounts in both well-lit and darkened trays (Fig. 2b). By comparison, when the jays cached in front of a conspecific, they cached significantly more pine nuts in the darkened sites (Fig. 2b). In the observed condition, caching in the well-lit sites was suppressed compared to the darkened sites whereas it was not in the in private condition. These findings suggest that the birds were actively avoiding caching in the well-lit sites when observed, but not when they cached in private.

At recovery, the jays retrieved very few items from the trays, however of those caches that were recovered; the vast majority were removed from the well-lit tray and re-cached in new sites (Fig. 2c). One suggestion is that this function as an additional cache protection strategy; cache some items in "risky" cache sites, such

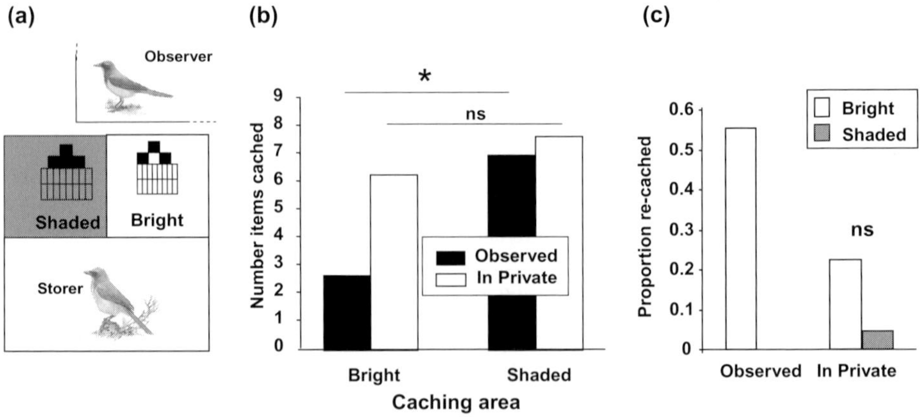

FIG. 2. Sun-shade experiment. **a** Schematic representation of experimental design. A unique caching tray is located in the shaded area of the cage and the bright area of the cage. **b** Number of food items cached in either bright or shaded trays; either when caching in private or when observed. **c** Proportion of food items re-cached from either bright or shaded sites; either when cached in private or when observed. (Adapted from Dally et al. 2004. *$p < 0.05$, ns $p > 0.05$)

as well-lit areas, possibly as a means of distraction, and then recover and re-hide those remaining items in the risky cache sites at a later time in private. These results cannot be readily explained by an aversion to cache in the light compared to the dark as the birds cache equal amounts in shady and well-lit areas in private (Dally et al. 2004). These results strongly suggest that western scrub-jays are sensitive to the risk of cache pilferage, preferentially caching in sites that will potentially degrade the quality of visual information available to pilferers and thus reduce the accuracy of observational spatial memory.

In a second experiment, scrub-jays were allowed to cache either in front of a conspecific or in private. In each condition, the scrub-jays were provided with two caching trays; one that was close to the observer's cage ("near tray") and a second that was relatively further away from the conspecific (i.e. at the back of the storer's cage "far tray"; Dally et al. 2005a; Fig. 3a). The suggestion being that it will be more difficult for a potential pilferer to steal caches made in sites further away either because they will focus their attention on those caches made closer to them or because the quality of visual information becomes degraded the further away a stimulus is located. When the scrub-jays cached in private, they cached equal amounts of wax worms in both the "near" and "far" trays. By comparison, when the storers cached in front of a conspecific they cached significantly more wax worms in the "far" tray than the "near" tray (Fig. 3b).

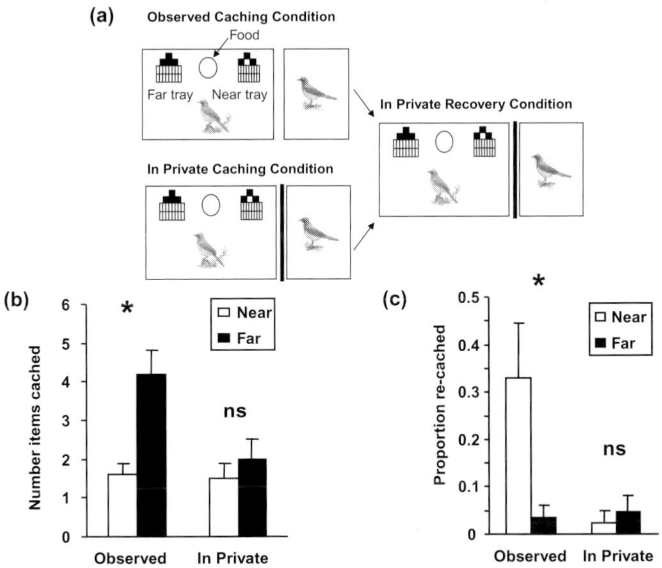

FIG. 3. Near-far experiment. **a** Schematic representation of experimental design. A unique caching tray is located close to a conspecific (near) and relatively further away (far). **b** Number of food items cached in either the near or far tray when cached in private or when observed. **c** Proportion of food items re-cached from either the near or far tray when the storer had cached previously in front of a conspecific (observed) or in private. (Adapted from Dally et al. 2005a, *$p < 0.05$, ns $p > 0.05$)

At recovery, the majority of items that were re-cached in new sites were from the "near" tray and only when they had been cached in front of another scrub-jay (Fig. 3c). Similarly to those caches made in the well-lit tray in the previous experiment (Dally et al. 2004), the caches made in the "near" tray would have been the most at risk of any pilfering attempts, as they were the closest to the potential pilferer, with the clearest view of exactly where they had been made. As with the Dally and colleagues (2004) experiment, the suggestion for why scrub-jays do not put all there caches in the most protected tray (e.g. "far" tray) is that caching a few items in the "near" tray may function as a form of distraction. As observers did not get the opportunity to steal any caches in this study, the storers were given the opportunity to retrieve these high risk caches. We will return to the issue of why the storers use cache protection strategies at all in the laboratory setting later.

In the final experiment investigating visual perspective taking, scrub-jays were provided again with two caching trays, but always with another bird present. In this study, one half of the storer's cage was occluded by a solid metal sheet. One caching tray was located behind the barrier which prevented the conspecific from observing any behaviour directed towards caches made in this "out of view" tray. A second tray was located in the other half of the storer's cage, but in full view of the conspecific, therefore the observer could see any behaviour directed towards caches made in this "in view" tray (Fig. 4a; Dally et al. 2005a). In this study, birds

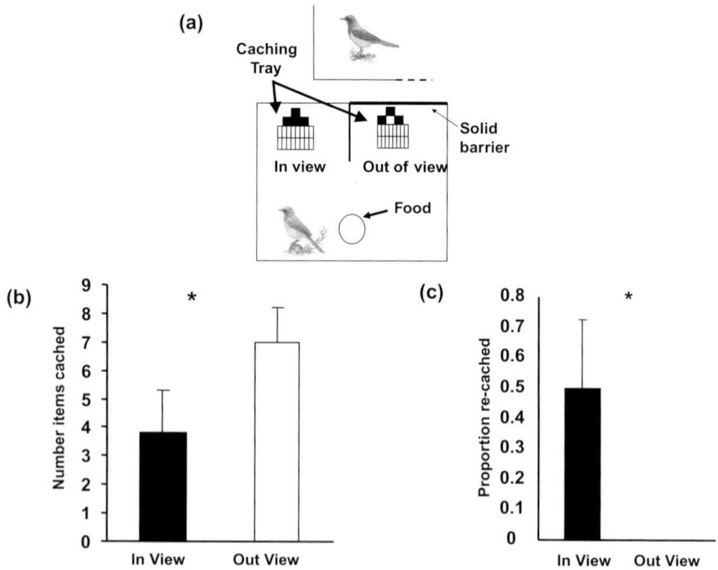

FIG. 4. Barrier experiment. **a** Schematic representation of experimental design. A unique caching tray is located "in view" and "out of view" (behind barrier) of a conspecific. **b** Number of food items cached; either in view or out of view. **c** Proportion of food items re-cached from either in view sites or out of view sites. (Adapted from Dally et al. 2005a, *$p < 0.05$, ns $p > 0.05$)

that either appreciate that another bird cannot see them if they cache behind a barrier, or prefer to cache out of sight of a conspecific, should choose to cache in the "out of sight" tray rather than the "in view" tray. Indeed, during caching, the storers preferentially hide caches in the "out of view" tray, however as before, they do place about 25% of their caches in the riskier "in view" tray (Fig. 4b).

At recovery, we found a similar pattern of results to the previous studies, in which the storers tended to re-cache items from the risky sites in the "in view" tray rather than the "out of view" tray (Fig. 4c). Therefore, in all three studies, the scrub-jays cached a few items in sites which the observers could see them clearly (well-lit, in the open or close) and the majority in the more protected sites. In every case, the storers returned to the riskier cache sites and re-cached those items in new places that the observer would have had no knowledge.

4.3. Knowledge Attribution[1]

The studies by Emery and Clayton (2001) and Dally et al. (2004, 2005a) suggest that storing scrub-jays may appreciate the different knowledge states of observers; if they had ⟨seen⟩ specific caches being made in tray A, they should be ⟨knowledgeable⟩ about the caches in A, whereas, if they did not ⟨see⟩ caches being made in tray B then they should be ⟨ignorant⟩ about the caches in B.

We therefore tested scrub-jays using a similar design to this (Dally et al. 2006a). A storer was allowed to cache in Tray A in the presence of Observer A (Fig. 5a). A second tray (B) was also present during caching, but the cache locations were not available as the tray was covered by a clear Plexiglas strip. After a short delay, the same storer was allowed to cache in Tray B (the Plexiglas strip had been removed) in front of Observer B, whilst Tray A was now covered (Fig. 5a). After a 3 hour period, both trays without covers were returned to the storer, and they were allowed to recover their caches in one of 4 conditions. They were either allowed to recover in the presence of Observer A, Observer B, a naïve bird who had not witnessed caching in either tray or in private (with no bird able to observe recovery).

During recovery, storers selectively recovered caches from those trays which had been observed during caching, when allowed to recover in front of those birds which had observed caches being made in those trays. For example, storers selectively re-cached items from Tray A when recovering in the presence of Observer A, but did not re-cache any items from Tray B (which Observer A had not witnessed being made, and therefore would not have any knowledge of these caches). Similarly, storers selectively re-cached items from Tray B when

[1]It should be noted that knowledge attribution has also been investigated in magpies (Prior and Güntürkün, unpublished observations) and ravens (Bugnyar and Heirich 2005), but we do not have the space to go into the details of these experiments. Although the results from both studies are extremely interesting, they do not possess the same levels of control as the scrub-jay experiments (Dally et al. 2006a), and consequently the results can be explained in terms of behaviour-reading, rather than mind-reading.

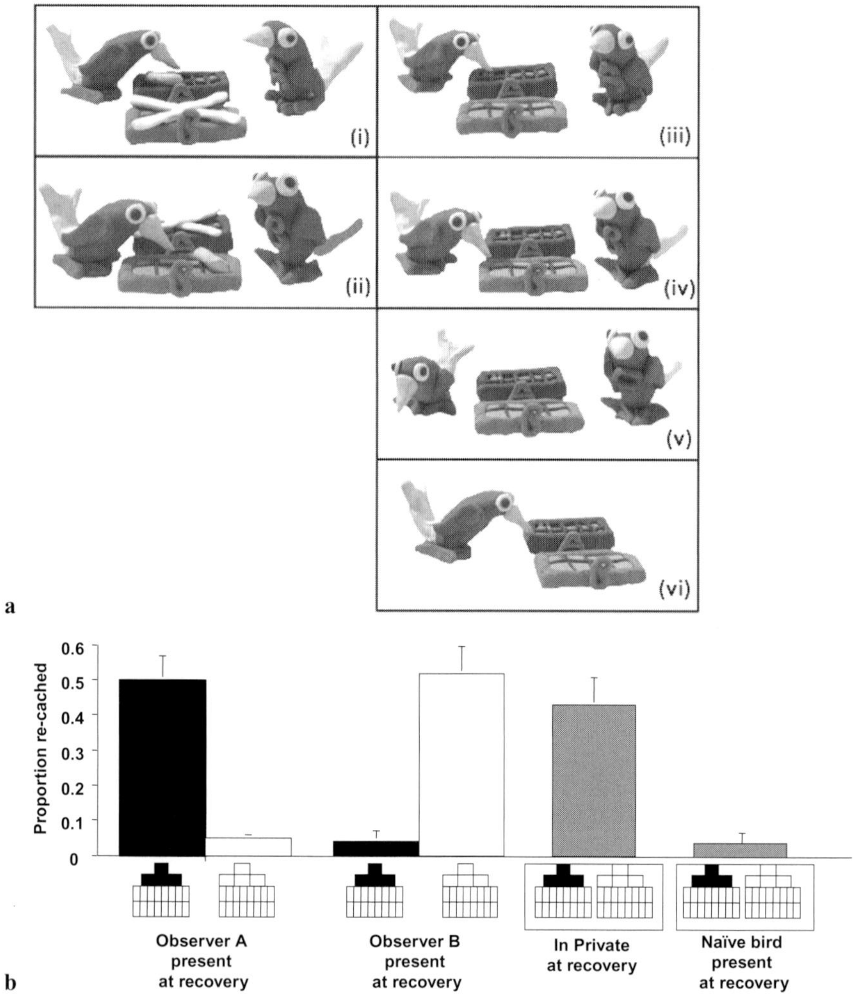

FIG. 5. Knowledge attribution experiment. **a** Schematic representation of experimental design. **i.** Storer caches in Tray A in front of Observer A, Tray B is covered with clear plastic; **ii.** Storer caches in Tray B in front of Observer B, Tray A is covered with clear plastic; **iii.** Storer recovers from Trays A and B in front of Observer A; **iv.** Storer recovers from Trays A and B in front of Observer B; **v.** Storer recovers from Trays A and B in front of a naïve conspecific (C) and **vi.** Storer recovers from Trays A and B in private. **b** Proportion of food items re-cached, either when in the presence of Observer A, Observer B, a naïve bird or in private. **c** Number of food items re-cached at recovery; **i.** re-cached from both trays into old and new sites during in private condition, **ii.** number of times food items moved in the in private condition, **iii.** re-cached from both trays into old and new sites during observed condition and **iv.** number of times food items moved in the observed condition. (Adapted from Dally et al. 2006a)

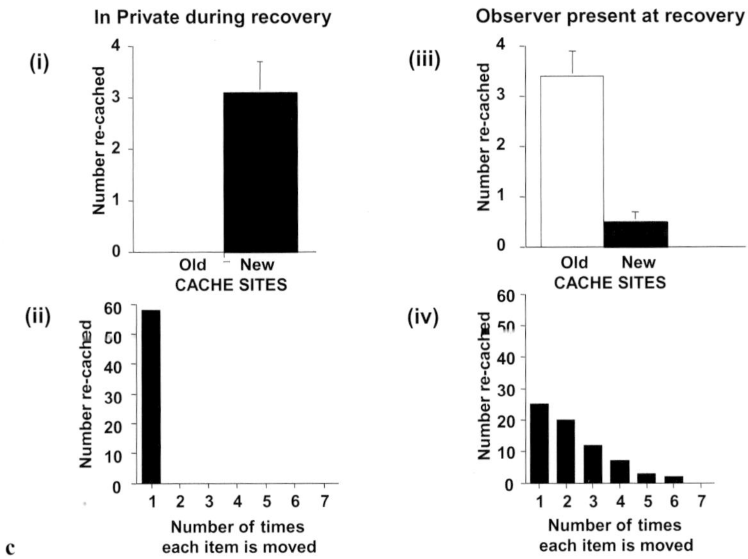

FIG. 5. *Continued.*

recovering in the presence of Observer B, but not from Tray A which Observer B had not witnessed being cached in (Fig. 5b).

By contrast, storers did not re-cache items from either tray when recovering in the presence of the naïve bird, suggesting that they attributed "ignorance" to that bird, which had not seen caches being made in either tray. Re-caching in front of the naïve bird would have provided them with information they previously did not have. Finally, the storers re-cached items from both trays when they recovered in private, as Observers A and B had seen caches being made in these trays, and so re-caching them in private would move the caches to new places which potential pilferers had not seen. This condition replicates earlier studies of re-caching in private when previously cached observed (Emery and Clayton 2001; Emery et al. 2004).

At first glance, re-caching in front of another bird does not seem like the most sensible strategy to protect caches. Indeed, dominant birds may not need to re-cache at all, and may defend their caches aggressively (Dally et al. 2005b). However, birds of a lower status should move their caches if they had been observed because of the threat provided by a pilferer with observational spatial memory. But, moving caches in front of another bird should provide new information on any new cache site. We therefore viewed the videotapes of recovery behaviour in detail, and noticed that the storers did not just move their caches only one time; they moved them up to 5–6 times during the short recovery period, but they only moved those caches which had been made in "observed trays" (i.e. Tray A in the presence of Observer A and Tray B in the presence of Observer B; Fig. 5c).

They moved caches once if they recovered in private (Fig. 5c), and not at all if they recovered in front of a naïve individual or from the "other tray" (i.e. not previously observed by Observer A or B). Most surprisingly, when the storers moved the caches multiple times, they did not necessarily leave the caches in the last location in which they made a bill probe (i.e. action used during caching). In fact, many caches were left in earlier locations, but the scrub-jays continued to a few times afterward, possibly "pretending" to cache or making "false" caches (Clayton et al. 2007). This behaviour may suggest a sophisticated level of tactical deception (similar to "creating a [false] image" as described by Whiten and Byrne 1988).

The basic results of this experiment can be interpreted in three ways; a low-level explanation based on associative learning, a mid-level explanation based on sophisticated behaviour-reading (e.g. purposive behaviour), but no understanding of mental states and a high-level explanation based on the attribution of knowledge. The low-level interpretation would suggest that the scrub-jays had associated the action of caching in a specific tray with the presence of a specific individual. At recovery, the presence of that specific individual would trigger the memory of having cached in that particular tray.

The mid-level explanation would suggest that the storers were in-tune with behavioural cues produced by the observers, such as spending more time attending to the tray they had cached in. As such, the storer was doing nothing more than reacting to the behaviour of the observer at the time of recovery. By contrast, the high-level interpretation would suggest that the storers not only remembered which particular individual was present during specific caching but also attributed different mental states to these individuals based on their presence ("knowledge") or absence ("ignorance"). The knowledge attribution would also need to be definitive to particular individuals in specific contexts, given that both observers were both knowledgeable of some caching events and ignorant of others. For example, Observer A had "knowledge" of the location of caches in Tray A, but was "ignorant" of caches in Tray B, and vice versa. The high-level interpretation may be reasonably convincing for corvids and apes, which have passed the higher level stages of object permanence (Bugnyar et al. 2007; Pollok et al. 2000; Tomasello and Call 1997).

However, what credence do these lower level interpretations have? Heyes (1998), for example, explained the results of earlier experiments on knowledge attribution in chimpanzees by Povinelli et al. (1990), as a form of associative learning; an individual was either present or absent during baiting and therefore simple associations could form based on presence ("knowledge") and absence ("ignorance"). This simpler explanation could also explain the selective re-caching behaviour of the scrub-jays, albeit it is more sophisticated as it has to be based on integrating information from the past (social context during caching) with planning for the future (potential pilfering by the observer). Note that the jays' behaviour cannot be explained in terms of differential familiarity with the two trays because both trays were always present during caching, although only one of them was available to cache in. Finally, no current associative theory explains such rapid one-trial learning, except in the context of taste aversion learning and since there was only one caching and recovery trial for each condition, the storers had no opportunity to be rewarded or punished for re-caching.

However, as we stated earlier, there is a third alternative interpretation, in which the storer's re-caching behaviour may have been guided by the behaviour of the observer rather than an explicit understanding of their unobservable mental state of knowledge. We therefore designed an experiment which attempted to rule out a "behaviour-reading" account, rather than an account based on reading observable mental states.

In this control experiment, Dally and colleagues presented the storer with two trays in which it could cache, either in Tray A in the presence of Observer A or in Tray B in the presence of Observer B, but in each case with the "other tray" covered over (Dally et al. 2006a). We then removed the trays and presented then to a second storer (Bird C), who was allowed to cache in either tray in the presence of an additional observer D (Fig. 6a). Then, at recovery, both trays were returned to the original storer and they were allowed to recover caches, either in the presence of the two original observers (A or B) or the new storer (D).

If the storer based their re-caching decisions on the behaviour of the observer bird, independent of whether or not they were present during caching in a particular tray, then the storer should have re-cached items equally from both the "observed tray" and the "other tray" because Observer D had witnessed caching in both trays (although by a different storer, C). Observer D should therefore have attended both trays equally, and so both trays would be equally at risk. By contrast, if the storer was attributing knowledge to the observer, they should, as shown previously, only re-cache items from the "observed tray". Indeed, the

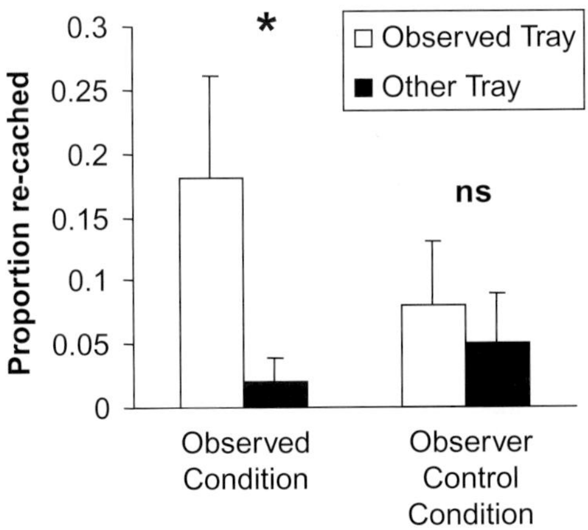

FIG. 6. Proportion of food items re-cached from either the observed tray or the other tray in the Observer condition or from either the observed tray or the other tray in the Behaviour-reading control condition. (Adapted from Dally et al. 2006a)

re-caching behaviour described earlier was repeated in the same observer condition (selectively re-cache from the "observed tray") whereas for the new behaviour-control condition (with Observer D), very little re-caching occurred, and what did was not significantly different between the two trays (Fig. 6b). We therefore suggest that it is unlikely that the storer was using behavioural cues from Observer D to influence their re-caching decisions.

4.4. Does It Matter Who's Watching?

One potential problem with the previous control experiment is the fact that the new observer (D) was completely new to the storer. There is abundant information which suggests that many birds, including corvids, can recognise individuals either by voice or by visual appearance. Therefore, it is likely that the storer recognised this particular individual, and remembered that they were not present at the time of caching, and therefore treated them similarly to the naïve bird in the original experiment (Dally et al. 2006a), and so not re-cache. Also, the observer might respond differently depending on whether the bird was the same as the one present at caching.

There are additional clues that scrub-jays recognise which individual was watching during caching based on the protective actions they utilise during cache recovery. Storers were allowed to cache in either a near tray or a far tray, either in front of a dominant conspecific, a subordinate conspecific, their partner or in private. The storers cached significantly more items in the far tray when caching in front of a conspecific, independent on which class of individual; dominant, subordinate or partner (Fig. 7a; Dally et al. 2006a). There were more caches made in front of the dominant individual than in front of the subordinate or partner.

At recovery 3 hr later, the storer re-cached a greater number of items from the "dominant trays" and specifically from the "near tray" (Fig. 7b). As with the earlier experiments on visual perspective-taking (Dally *et al.* 2004, 2005a), the scrub-jays cached some items in risky sites, such as near to the observer. In the current experiment, items were re-cached specifically from the "near tray" in front of a dominant. This may be explained by the fact that dominant birds aggressively steal the caches of subordinates, whereas subordinates have to resort to more subtle cache protection strategies, and partners cache readily in front of one another and share their caches (Dally et al. 2005b).

4.5. Caching & Pilfering as A Cognitive "Arms Race"

During a caching episode, cachers and pilferers may utilise many different strategies to either protect their caches or increase the possibility to steal caches without the subsequent threat of aggression from the storer. As yet, we do not know what strategies pilferers use to increase the chances with which they will successfully steal another's caches, but we suggest that they will be similar to those used by storers. For example, we suggest that it will aid a thief to hide

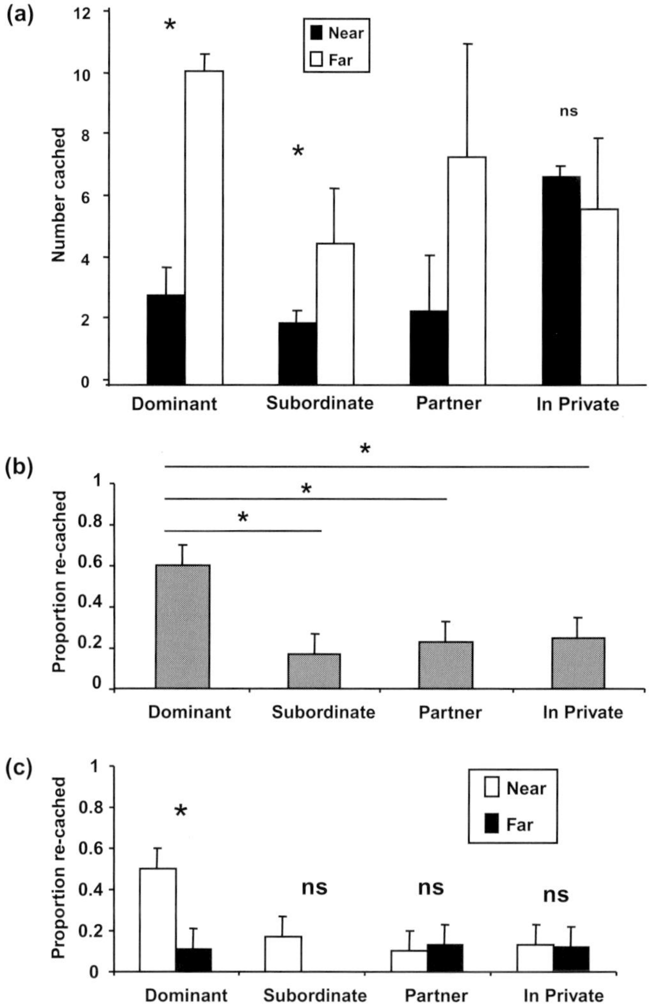

FIG. 7. Observer identity experiment. **a** Number of food items cached in either the near or far tray in front of a dominant, subordinate or partner bird or in private. **b** Proportion of food items re-cached in private at recovery when they had cached previously in front of a dominant, subordinate or partner bird or in private. **c** Proportion of food items re-cached in private from either the near or far tray when they had cached previously in front of a dominant, subordinate or partner bird or in private. (Adapted from Dally et al. 2006a. *$p < 0.05$, ns $p > 0.05$)

themselves during caching; either at a distance, in the shade or behind a barrier, such as a tree or rock.

It is not clear how such strategies and counter strategies may have evolved, however, the use of such strategies in a flexible manner, as described for the previous experiments, suggests complex cognition and a high degree of innovation. Indeed, the fact that cache protection strategies are not employed until a storer has had the specific previous experience of stealing another bird's caches reinforces this idea (Emery and Clayton 2001, 2004b).

The fact that the same individual can play both the role of cacher *and* pilferer is important for the relationship between cache protection and mind-reading, as an individual needs to not only keep track of it's own role (cacher or pilferer), but also the various cognitive strategies employed to either protect caches or steal other's caches (Clayton et al. 2007; Dally et al. 2006b). We suggest that such a "cognitive arm's race" will (a) increase the frequency of novel social innovations (as new protection strategies will require new pilfering strategies, and vice versa), (b) that these innovations will become transmitted relatively quickly through a group (if they are successful and the species involved lives in a social group), and (c) will likely be based on a psychological system which allows individuals to "put themselves into another's shoes" (discussed below).

The problem with having to invent new strategies is that if they are successful, and pass quickly through a group, then the necessity for inventing new strategies increases at the same rate. This scenario is not practical; therefore it is likely that existing strategies will develop into more sophisticated forms that become less predictable, such as moving caches around multiple times. With this particular strategy, it is never clear where the caches will end up, and so this strategy can be used frequently without reducing its impact on pilfering efficiency. The same cannot be said for examples of tactical deception described for primates (Whiten and Byrne 1988). Unfortunately, the use of this strategy comes at a price; the more the caches are moved around, the greater the potential for interference of the storer's memory for their own caches, and the increased amount of time taken to move a single cache around multiple times decreases the amount of time available to protect the other caches and increases the likelihood that the observer might see the cache.

4.6. Experience Projection

The cache protection strategies described above suggest that storers have not only equipped themselves with a vast array of different cognitive tools to counter the potential for pilfering, but also that these abilities may require the storer to place themselves "inside the head" of a pilferer and use the specific counter measures to counter what they would do as a pilferer in the same situation. This in itself suggests sophisticated abilities hitherto not described for other non-human animals. These abilities imply something which has been termed *experience projection*. Humphrey (1980) was perhaps the first to suggest that inward reflection or introspection was an essential part of understanding other social beings.

> "a revolutionary advance in the evolution of mind occurred when, for certain social animals, a new set of heuristic principles was devised to cope with the pressing need to model ... the behaviour of other kindred animals. The trick which Nature came up with was introspection ... it proved possible for an individual to develop a model of the behav-

iour of others by reasoning by analogy from his own case ... The task of modelling behaviour does indeed demand formidable intellectual skill ... but intelligence alone is not enough. If a social animal is to become ... one of **"Nature's psychologists"** it must somehow come up with the appropriate ideology for doing psychology" (Humphrey 1980, p. 59).

Humphrey (1980) suggested that social animals, and he did not explicitly state which ones, could use their own experiences and perspective to model another's thoughts in the same situations, and so predict their actions.[2] Gallup (1982) suggested something similar which has been named "experience projection", in which an individual can use their own experiences to infer experiences in others, and so forming the foundation for a theory of mind (simulation and empathy). Indeed, Gallup stated that if an individual could represent mental states in others it should, by necessity, be aware that their self is different from another's self. As such, self-awareness, mind-reading and consciousness are all inter-linked, and evidence for one is highly indicative of evidence for all (Gallup 1982).

Gallup (1982) proposed a paradigm which could test for this in animals (see also Heyes 1998; Povinelli and Vonk 2003). A chimpanzee is given experience wearing a blindfold or goggles which are either opaque or transparent, but the only way of telling which is which is by the addition of an arbitrary cue, such as coloured rims on the goggles. The chimp will have never seen or worn these goggles before. After a period in which the chimp gets to wear these two types of goggles, the chimp is then allowed to interact with another chimp who is wearing similar coloured goggles (opaque or transparent). Will the chimp adjust their normal behaviour with respect to the chimp based on a) their previous experience wearing the goggles and the obstruction of their view and b) whether the chimp is wearing the opaque or transparent goggles? This experiment has yet to be done (although Povinelli and Vonk 2006 briefly discuss a similar design using buckets, but no results were presented).

An experiment in scrub-jays is perhaps the only example of a test for experience projection in animals. Recall an earlier experiment, in which scrub-jays cached either in front of a conspecific ("observed") or "in private", and at recovery (always in private) the birds who had been observed previously re-cached food items primarily into new sites, whereas when they had cached in private they performed very little re-caching and did not differentiate between old and new sites (Emery and Clayton 2001). Before this study, the birds were tested for observational spatial memory and split into three groups (Clayton et al. 2001). One group were allowed to cache wax worms in trays (Storer group) and later recover them in private, a second group watched the storer cache and were then allowed to find their caches and steal them (Observer group) and a third group were located in an adjacent cage to the caching bird, but the view between them was occluded, so they could only hear them caching (Control group). The birds

[2]Clayton et al. (2007) argue that the western scrub-jay falls into Humphrey's category of being one of "Nature"s psychologists.'

in the Observer group were almost as accurate in locating caches as the birds in the Storer group, whereas the Control group birds were poor at finding caches (Clayton et al. 2001).

Importantly for the Emery and Clayton (2001) study, these three groups had different experiences based on caching and pilfering. The Storer group (Clayton et al. 2001) had experience of caching and recovery, and had seen other birds caching, but had never been given the opportunity to pilfer another jay's caches. Consequently, this group was renamed the Observer group in the Emery and Clayton (2001) study. The Observer group in the Clayton et al. (2001) study had experience of observing other's cache, and had been given the opportunity to steal those caches, and so was re-named the Observer + Pilfer group (Emery and Clayton 2001). Finally, the Control group in the Clayton et al. (2001) study had experience of pilfering other birds caches, even though they had not seen them being made, and so were re-named the Pilfer group (Emery and Clayton 2001). We do not suggest that the Pilfer group had never seen another bird caching, only that in these particular studies they had not been given any specific experiences of observing and pilfering the *same* caches.

We were interested in whether these different experiences would lead to differences in the presence or absence in the use of cache protection strategies. Indeed, the Observer + Pilfer, and the Pilfer groups demonstrated an identical pattern of results; re-caching a large proportion of items from the tray in which they had previously been observed caching, and moving them to new sites (Fig. 8a,b). By contrast, the Observer group without any experience of being a

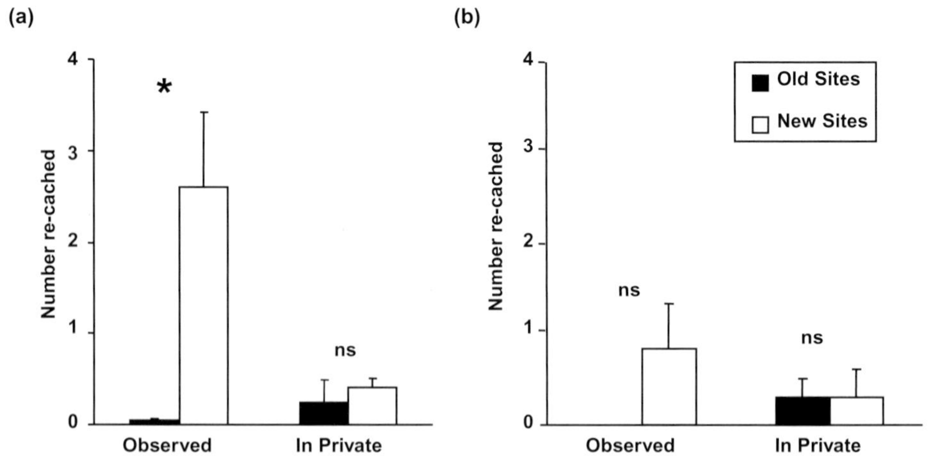

FIG. 8. **a** Number of food items re-cached by Pilfer group into old or new sites when they had been previously observed during caching or had cached in private. **b** Number of food items re-cached by Observer group into old or new sites when they had previously been observed during caching or had cached in private. (Adapted from Emery and Clayton 2001)

pilferer themselves re-cached very few, if any items (Emery and Clayton 2001; Fig. 8c). Therefore, we suggest that storers with the specific experience of stealing another's caches (even when they had not seen the caches being made), *project* this experience onto another bird, namely one which is observing their caching, and so is a potential thief.

Therefore, during recovery the storers with pilfering experience have to (a) remember the previous social context during caching (i.e. whether they were observed or in private), (b) reflect on their previous pilfering experience (i.e. what they did when observing and pilfering), and (c) use this experience to guide their decisions about what to do with their caches when no observer is present (i.e. re-cache in new sites which the observer has no recourse to search in).

It is not clear how this pilfering experience may be represented in the scrub-jay brain (however a proposal will be discussed in the final section). This issue is particularly intriguing with respect to the Pilfer group. In the observational memory study (Clayton et al. 2001), these birds did not see another bird caching and thus only pilfered caches they had not seen being made. How did these birds ⟨know⟩ that they were pilfering another's caches and not recovering their own without some understanding of self versus other? The birds did have experience of hearing another bird cache, although it would have been difficult to differentiate the sounds of caching from recovery. They also may have a natural tendency to search opportunistically for caches. However, these two factors do not explain how the birds could differentiate between caches they had made and caches others had made.

One might suggest that all these birds had to do was remember that they did not cache in these particular trays (as each tray was unique). Although these birds did not have experience of stealing caches made by other birds, they had considerable experience of caching and recovery from similar looking trays, so it likely that they had generalised that all trays which resemble trays in which they had previously cached in, also contain caches. However, to translate the action of removing caches not made by them into the specific experience of being a pilferer, and the subsequent action of re-caching in new sites when they had previously caching in front of a conspecific, suggests more than this simple explanation. It is hard to reconcile without some form of understanding about own versus other.

We can be fairly certain that it was the act of being a pilferer rather than being pilfered from which was the relevant experience, as all birds in these groups will have been pilfered from, as they had all taken part in studies of episodic-like memory (e.g. Clayton and Dickinson 1998) in which all caches were removed by the experimenters before recovery to test for memory. Only the Observer + Pilfer group and the Pilfer group had had pilfering experience.

4.7. But Is It "Theory of Mind"?

A sceptic might be quick to say that what the scrub-jays are doing is either sophisticated behaviour-reading in the case of cache protection strategies (as the observer is present during caching), or they are behaving based on a series of species-specific predispositions which enable them to become "in tune" with a pilferer's actions because this is biologically important (i.e. is a domain-specific form of cognition).

Although no one has yet to criticise the scrub-jay (or raven) social cognition studies in print, we would like to attempt to do so ourselves along the same lines as Povinelli (2000, Povinelli and Vonk 2003, 2006) and Heyes (1998) have done for chimpanzee social cognition studies, particularly those based on visual perspective-taking. We believe that this exercise will be useful in terms of collating evidence for theory of mind in scrub-jays, helping in the design of future experiments which might be less ambiguous in their interpretation, but also revealing whether alternative explanations are not sufficient to explain the scrub-jays' behaviour.

Povinelli's main criticism of the Leipzig group studies (e.g. Hare *et al.* 2000, 2001) is that the chimpanzees' behaviour (moving toward the hidden food) can be explained by a psychological system which represents another's behaviour (S_b) and by a system which represents another's behaviour plus their mental states (S_{b+ms}). Povinelli rightly suggests that folk psychology is constructed from an understanding of behaviour; therefore it is difficult to distinguish between the low-level S_b explanation and the high-level S_{b+ms} explanation (Povinelli and Vonk 2006). This argument is particularly cogent for the studies of visual perspective-taking, as the chimps can both see one another, and even when they cannot see one another at the start of the test, the subordinate will have seen the facial orientation of the dominant chimp, probably looking at the piece of food that both of them can. ⟨see⟩ The subordinate may therefore have avoided the piece they both could see and approached the other piece because the dominant was oriented towards the piece in the open, not because the subordinate knew that this was the piece of food that the dominant could ⟨see⟩.

Using Povinelli's methods, the mental states which might be involved in the Hare et al. (2000) study may be explained as follows:

(a) The subordinate chimp observes both pieces of food, one located next to a barrier, one located in the open;
(b) The dominant chimp observes one piece of food located in the open;
(c) The subordinate chimp concludes that because the food next to the barrier is hidden from the dominant's view, they cannot ⟨see⟩ it, whereas the food in the open can be ⟨seen⟩ by both chimps;
(d) The subordinate chimp approaches the food next to the barrier (because if the dominant could ⟨see⟩ the hidden food it would move towards and compete for it).

Section (c) tries to explain the subordinate's behaviour based on reasoning about mental states (⟨seeing⟩), however Povinelli suggests that removing this explanation does not affect the behavioural outcome of the experiment (i.e. that the subordinate still approaches the hidden food).

(a) The subordinate chimp observes both pieces of food, one located next to a barrier, one located in the open;
(b) The dominant chimp observes one piece of food located in the open;
(c) The subordinate chimp approaches the food next to the barrier (because chimps only approach food that is available to them).

In this case, an explanation that is not based on representing mental states is easy to reconcile, as chimps will have had many previous occasions (outside of the experiments) in which a dominant will get access to food when presented, but will not get food that is not within their field of view. This analysis falls down, however, when learning and experience cannot account for the behaviour (see later). What of the scrub-jays? Using the same methodology, we will describe the various visual perspective-taking experiments, and determine whether S_b has the same explanatory power as S_{b+ms}.

Let us consider the barrier experiment (Dally et al. 2005a), in which the scrub-jay either cached behind a barrier or in the open. Povinelli's description may go as follows:

(a) The storer is in the presence of an observer when the tray is in the open;
(b) The storer is alone when the tray is located behind a barrier;
(c) The storer concludes that because the tray is hidden behind a barrier, the observer cannot ⟨see⟩ the contents of the tray;
(d) The storer caches in the tray behind the barrier (because the observer cannot pilfer caches that it cannot ⟨see⟩).

In this example, the explanation based on reasoning about mental states (S_{b+ms}) does not do any additional causal work than the explanation based on reasoning about behaviour (S_b). The observer is either absent (when the storer caches behind the barrier) or present (when the storer caches in the open), therefore the decision of where to cache may be dictated by two conditional rules; "out-of-sight, out-of-mind", and "others are unlikely to steal caches if they were absent when the caches were hidden or caches are less at risk if they are hidden in private, behind barriers, in dark places, etc.":

(a) The storer is in the presence of an observer when the tray is in the open;
(b) The storer is alone when the tray is located behind a barrier;
(c) The storer caches in the tray behind the barrier as another bird is not present (and others do not steal caches when they were absent during caching).

In the wild, corvids tend to cache next to objects, such as rocks, tree stumps and changes in patches of grass (Bennett 1993; Bugnyar and Kotrschal 2002; Dally et al. 2006b; Helme et al. 2004, unpublished observations), however it is not clear whether this is because they act as landmarks allowing the caches to be easily found again by the storer, or as a form of protection against conspecifics seeing where the caches have been made (de Kort et al. 2006). In the Dally et al. (2005a) study, only one trial was provided for each condition, and consequently they had little opportunity to learn that caching alone results in little or no pilfering, whereas caching in front of others results in caches being stolen, particularly since these hand-raised birds had not had any prior experience with barriers. Nonetheless, the possibility remains that such rules are innate, or rapidly generalised from the previous social context of caching experiments (Clayton et al. 2001; Emery and Clayton 2001).

A more difficult case to explain by reasoning about behaviour alone is the sun/shade experiment (Dally et al. 2004). In that experiment, a storer was provided with 2 caching trays, one located in the dark and the other in the light. A Povinellian analysis would go along the following lines:

(a) The storer is in the presence of an observer with two caching trays; one in bright light, one in the dark;
(b) The tray in the light is easier to ⟨see⟩ than the tray in the dark, and so easier to pilfer from;
(c) The storer caches in the dark tray (because the observer cannot ⟨see⟩ the caches as clearly as those made in the bright tray).

However, if we remove the mental state reasoning from (b), then we are left with the following:

(a) The storer is in the presence of an observer with two caching trays; one in bright light, one in the dark;
(b) The tray in the light is more visible;
(c) The storer caches in the dark tray (because observers do not pilfer as successfully from dark trays).

A simple "out of sight, out of mind" discrimination of the type described for the barrier experiment does not work in this case, as the storer cached in both trays in the presence of another bird, and it was the trays which were different not the observers. Indeed, the only difference between the two trays (besides being visuo-spatially unique) was the amount of light projected upon them. One possibility is that the storers use a conditional rule based on an assessment of cache risk in which the observer's behaviour differed depending on whether the storer was caching in the well-lit tray or the dark tray.

It is not clear how this difference was manifest. Aside from an innate conditional rule, it is possible to conjure up a behaviour-reading account. For example, it is possible that the observer looked more closely at the caches made in the dark tray because it would be more difficult to see where the caches had been made (i.e. our functional explanation for why the storer cached more items in the dark tray). If the observer looked more closely, then we predict that the storer should have avoided caching in this tray, not increasing the amount, as was found. Of course, the opposite may have occurred where the observer could not have seen the caches being made in the dark tray and so lost interest and focused their attention onto the light tray, and consequently the storer avoided caching in the light tray. The storer did make most of their caches in the dark tray; however it seems unlikely that the observer would avoid looking at the tray in which the majority of the caches were made, if they were to use observational spatial memory to pilfer them at a later time.

Behavioural explanations of this experiment should also appreciate that the storer needs to use a second conditional rule, namely one about their own position with respect to the two trays and the observer, as both trays are present in both conditions ("observed" and "in private"), and it is the differential environment of the storer (in the dark or in the light) which predicts their behaviour.

The one problem with the behaviour-reading account is that it requires a series of very flexible conditional rules which can be applied to different individuals across a variety of different but particular contexts. For example, "Public Enemy Number One" is one of the more dominant scrub jays in our colony who always looks out for pilfering opportunities, often using aggression to steal caches. As an observer, "Public Enemy Number One" looks intently at the caches being made and does not attempt to hide his "looking behaviour". By contrast, "Jack the Lad" is a subordinate scrub-jay who will pilfer when given the opportunity, but only when allowed to pilfer in private when the cacher has left the scene. As such, "Jack the Lad" has developed good observational spatial memory.

When "Jack the Lad" is cast as an observer, he tends to turn away from the cacher, or hide, particularly if the cacher is dominant, so as to avoid giving away his "pilfering intentions". Therefore, two different birds, two different pilfering strategies, and two different forms of behaviour which need to be read and interpreted based on identity and context. Furthermore, when "Sweetie Pie" caches then "Jack the Lad" is dominant, rather than subordinate, so he may look at the caches being made and revert to an aggressive strategy.

The behaviour-reading account also does not allow one to accurately predict the future behaviour of an unknown conspecific, as behaviour-reading is based on computing statistical regularities over the course of a relationship between two individuals. As western scrub-jays are semi-territorial, and cache protection may be primarily implemented to protect against theft by territory neighbours, it seems unlikely that such an unspecific method would be productive in preventing pilfering. In this case, understanding something about mental states, or being sensitive to complex conditional rules of the form described above, would appear to be the most biologically parsimonious explanation for their behaviour.

How does a Povinellian analysis stand up to the other experiments designed to test for epistemic states, such as ⟨knowledge⟩? To recap, in the Dally et al. (2006a) study, storers were presented with two trays (A & B) only one of which they could cache in (because the other was blocked access), in the presence of a specific observer (A or B). Therefore, they could cache in Tray A in the presence of Observer A, and cache in Tray B in the presence of Observer B. At recovery, both trays were available and storers could recover in the presence of Observer A or B, a naïve bird who had not been present at the time of caching or recover when alone. A Povinellian analysis may proceed as follows:

(a) The storer caches in Tray A in the presence of Observer A and access to Tray B is blocked;

(b) The storer then caches in Tray B in the presence of Observer B and access to Tray A is blocked;
(c) The storer concludes that only individuals who had ⟨seen⟩ caches being made will have ⟨knowledge⟩ of them;
(d) At recovery, the storer re-caches selectively from Tray A when in the presence of Observer A, and from Tray B when in the presence of Observer B because the observers have ⟨seen⟩ caches being made in these trays and so ⟨know⟩ their location. The storers do not re-cache from the other tray in the presence of the observers because they have not ⟨seen⟩ caches being made in those trays, and so will be ⟨ignorant⟩ of their location. The storers do not re-cache from either tray in the presence of an unknown observer as they have not ⟨seen⟩ any caching and so will be ⟨ignorant⟩ of all caches.

Again, removing the explanation based on reasoning about mental states (c) leads to the following revised explanation based solely on reading behaviour:

(a) The storer caches in Tray A in the presence of Observer A and access to Tray B is blocked;
(b) The storer then caches in Tray B in the presence of Observer B and access to Tray A is blocked;
(c) At recovery, the storer re-caches selectively from Tray A when in the presence of Observer A, and from Tray B when in the presence of Observer B because the observers directed behaviour [e.g. looking] towards the tray that was cached in when they were present or the presence of the observer triggered a memory of where they cached in the presence of the specific observer. The storers do not re-cache from either tray in the presence of an unknown observer as they were absent during caching, and pilfering does not occur in the absence of an observer during caching.

Although this low-level account may explain the action of re-caching (however, see below), it fails to explain the fact that the storers move caches around multiple times during recovery, and particularly when the caches are not placed in the final location in which the storer inserted their bill. We have already provided a functional explanation for this behaviour, which is that it aids to confuse the observer as to the new location of the caches; however it is not clear that a S_b is sufficient to explain the mechanisms involved. Perhaps, the storer waits to deposit the cache in a site when the observer is not looking or is distracted, and this influences the final cache site? This suggests a level of behaviour-reading so far not reported for any non-human animal. However, as unlikely as this scenario may be, the fact that it is possible presents a problem for our interpretation of this behaviour.

One argument against this suggestion is the result of our control experiment (Dally et al. 2006a). At recovery, the storer re-cached items from the "observed tray" when in the presence of a previous observer, but not from the "other tray", thereby replicating our previous experiment. In the observer control condition, in which a new observer is present when new caches are made by a new storer, the original storer does not re-cache many items, and does not differentiate between the two trays when in the presence of the new observer. We suggested that the storer rightly did not re-cache many items as they would have "recognised" that the new observer had not been present at caching. However, if the new observer looked more intently at one of the trays (i.e. the one they had witnessed caching in), then this should have been perceived by the storer. The storers did not appear to perceive such differences between the two trays, as there was no difference in the number of items re-cached from either tray. It remains to be determined which psychological process or processes are involved in this suite of protective behaviours.

For the final study to be considered (Emery and Clayton 2001), one might suggest that re-caching, even when performed in private, is elicited by a memory of the previous behaviour of the observing bird. This is difficult to reconcile with the results of an interleaved trials experiment in which the storers first cached in Tray A when observed, and then a short time later in Tray B when in private or vice versa (Emery and Clayton 2001). In this case, the storers displayed the same pattern of results at recovery as in the basic design; re-caching specifically from the tray in which they had cached when observed, and primarily in new sites. It is therefore hard to explain re-caching based on a general memory for the previous event or at a simpler level, a hormonal or other physiological stress response produced by being watched during caching. Indeed, if a simpler mechanism elicited re-caching it might predict generalised re-caching, not re-caching specifically in new places.

What level of explanation would be suffice for the fact that re-caching behaviour is not an automatic response, but is dependent on the storer having had previous experience of being a thief in the past? Let us try to reverse engineer the possible psychological mechanisms which may be involved at recovery to elicit re-caching.

First, the storer has to remember the previous social context of a previous caching event in relation to each tray (either bird present or absent). Second, the storer needs to use this social information to influence its current behaviour (re-caching) in anticipation of a future event (pilfering) independent of the present context (no observer present). Third, only jays with experience of being a pilferer transfer this experience to another bird, and so produce appropriate counter strategies (re-caching in new sites). Fourth, the idea that birds project their experience onto another bird may suggest a form of introspection. It is difficult to envisage how this could occur without recourse to some appreciation of the distinction between self and others.

5. Cognitive Architecture of the Scrub-jay Social Mind

The preceding analysis of the scrub-jays' behaviour in each of our experiments infers that their cache protection strategies are (a) rational (i.e. caused by the interaction of a belief and desire in such a way that performance of the behaviour in question fulfils the desire if the belief is true (and fails to do so if the belief is false; Clayton et al. 2006), (b) follows a logical progression, (c) based on a sophisticated understanding of the subtle behaviour of others which allow storers to predict the future actions of conspecifics, (d) based on the specific experience of being a pilferer and (e) maybe based on introspection.

If we conclude that western scrub-jays' cache protection strategies are rational, and therefore cognitive, we can use this information to construct a cognitive architecture of how scrub-jays do their psychology when competing with potential thieves (Fig. 9). We therefore present a cognitive architecture of the scrub-jay social mind inspired by a cognitive architecture of human mind-reading (Nichols and Stich 2003).

During recovery, a number of processing stages need to occur before a cache may be protected from pilferage. Information enters the PERCEPTION system to allow an assessment of the current social context; either the storer is alone or in the presence of other individuals. If the storer is in the presence of conspecifics, then their behaviour will also be assessed (e.g. eyes open or closed, direction of attention, direction of movement, vocalizations, emotional state, etc.). This information feeds into the BELIEF system, which alerts the bird to the possibility of theft; "*caches will be stolen by an observer who saw then being made*". The BELIEF system interacts with the DESIRE system which functions to motivate their response; "*protect caches from theft*". Information feeds into the DESIRE system from the INTERNAL ENVIRONMENT (e.g. motivational state). Information from the BELIEF and DESIRE systems passes to the DECISION

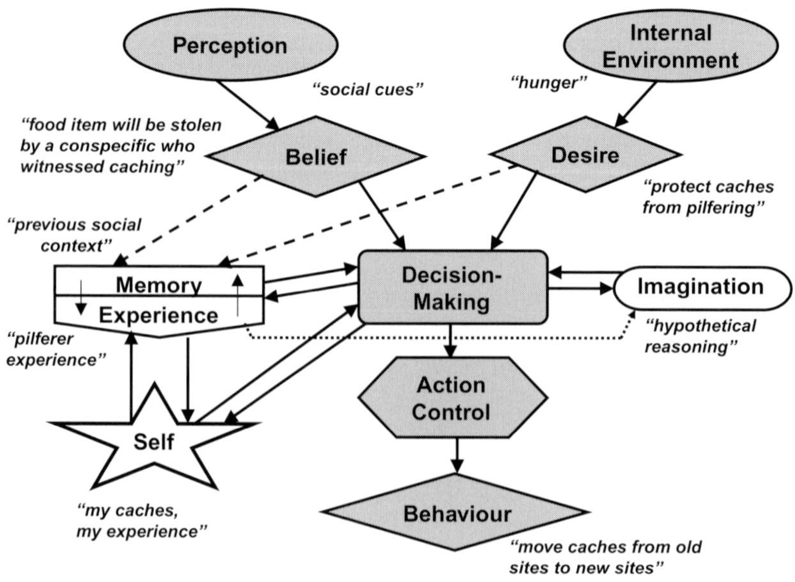

FIG. 9. Cognitive architecture of the scrub-jay social mind.

MAKING system, where the ultimate course of action (e.g. appropriate cache protection strategy) will be initiated. The DECISION MAKING system interacts with a number of different systems in order to make the correct decision. Alternative courses of actions may be rehearsed in the IMAGINATION system, allowing a storer to predict which actions will be expected to succeed, and which ones will fail. This system interacts with the MEMORY/EXPERIENCE system to provide information on which actions have been successful in the past. The MEMORY/EXPERIENCE system interacts with the BELIEF system as only thieves recognise the threat posed by conspecifics for cache safety. The MEMORY/EXPERIENCE system also provides information on the previous social context (either alone or in the presence of others). The MEMORY/EXPERIENCE system may also interact with a SELF system to determine that the experiences are "owned" by the storer and are not "possessed" by others. Once this information has been assimilated, and an appropriate course of action "decided" upon, the DECISION MAKING system communicates this decision to the ACTION CONTROL system, which initiates the appropriate behaviour (e.g. re-caching in new sites, moving caches around multiple times, hiding food in the shade, etc.).

6. Conclusions

We suggest that our detailed analysis of the different cache protection strategies used by western scrub-jays, and especially whether they require an understanding of behaviour alone or behaviour plus mental states, is good evidence for theory of mind in scrub-jays. We are particularly confident that we have provided good evidence based on Heyes' (1994) proposal for triangulation of evidence, and the important role of experience projection which is difficult to explain using behaviour-reading and simple associative theories of learning. Currently, studies on other non-human animals may be explained by behaviour-reading rather than understanding unobservable mental states. Of course, we do not propose that scrub-jays are particularly intelligent compared with other non-human animals, or that other animals do not have a theory of mind, only that the cache protection paradigm (and implementation of the 3Es) has provided us with an extremely useful tool with which to investigate these issues in non-verbal creatures based on their natural history. This ability to increase the ecological validity of experiments, although dismissed as unimportant by some (e.g. Povinelli and Vonk 2004), is an incredibly powerful weapon in our quest to learn more about the minds of other animals.

Acknowledgements. The writing of this chapter was supported by a Royal Society University Research Fellowship to NJE and grants from the BBSRC, Royal Society and University of Cambridge. We would like to thank Shoji Itakura and Kaz Fujita for their patience in waiting longer than humanly possible for this chapter to finally make an appearance.

References

Avian Brain Nomenclature Consortium (2005) Avian brains and a new understanding of vertebrate brain evolution. Nat Rev Neurosci 6:151–159

Balda RP (2002) Pinyon jay (*Gymnorhinus cuanocephalus*). In: Poole A, Gill F (eds) The birds of North America, No. 605. The Birds of North America, Inc, Philadelphia, PA

Balda RP, Kamil AC, Bednekoff PA (1996) Predicting cognitive capacity from natural history: examples from four species of corvids. In: Nolan V, Ketterson ED (eds) Current ornithology. Plenum Press, New York, pp 33–66

Bednekoff PA, Balda RP (1996a) Observational spatial memory in Clark's nutcrackers and Mexican jays. Anim Behav 52:833–839

Bednekoff PA, Balda RP (1996b) Social caching and observational spatial memory in pinyon jays. Behaviour 133:807–826

Bennett ATD (1993) Spatial memory in a food storing corvid: I near tall landmarks are primarily used. J Comp Physiol A 173:193–207

Bitterman ME (1975) The comparative analysis of learning. Science 188:699–709

Bond AB, Kamil AC, Balda RP (2003) Social complexity and transitive inference in corvids. Anim Behav 65:479–487

Bugnyar T, Heinrich B (2004) Ravens, *Corvus corax*, differentiate between knowledgeable and ignorant conspecifics. Proc Roy Soc B 272:1641–1646

Bugnyar T, Kotrschal K (2002) Observational learning and the raiding of food caches in ravens, *Corvus corax*: is it "tactical" deception? Anim Behav 64:185–195

Bugnyar T, Stöwe M, Heinrich B (2007) The ontogeny of caching in ravens, *Corvus corax*. Anim Behav (in press)

Clayton NS, Dickinson A (1998) Episodic-memory during cache recovery by scrub-jays. Nature 395:272–274

Clayton NS, Emery NJ (2004) Cache robbing. In: Bekoff M (ed) Encyclopaedia of animal behaviour. Greenwood Publishing Group, Westport, CT, pp 251–252

Clayton NS, Griffiths DP, Emery NJ, Dickinson A (2001) Elements of episodic-like memory in animals. Phil Trans R Soc B 356:1483–1491

Clayton NS, Emery NJ, Dickinson A (2006) The rationality of animal memory: the cognition of caching. In: Nudds M, Hurley S (eds) Rational animals? Oxford University Press, Oxford, pp 197–216

Clayton NS, Dally JM, Emery NJ (2007) Social cognition in food-caching corvids: the western scrub-jay as a natural psychologist. Phil Trans R Soc B 362:507–522

Dally JM, Emery NJ, Clayton NS (2004) Cache protection strategies by western scrub-jays (*Aphelocoma californica*): hiding food in the shade. Proc R Soc Biol Lett 271: S387–S390

Dally JM, Emery NJ, Clayton NS (2005a) Cache protection strategies by western scrub-jays: implications for social cognition. Anim Behav 70:1251–1263

Dally JM, Emery NJ, Clayton NS (2005b) The social suppression of caching by western scrub-jays (*Aphelocoma californica*). Behaviour 142:961–977

Dally JM, Emery NJ, Clayton NS (2006a) Food-caching scrub-jays keep track of who was watching when. Science 312:1662–1665

Dally JM, Clayton NS, Emery NJ (2006b) The behaviour and evolution of cache protection and pilferage. Anim Behav 72:13–23

Emery NJ (2000) The eyes have it: the neuroethology, evolution and function of social gaze. Neurosci Biobehav Rev 24:581–604

Emery NJ (2005) The evolution of social cognition. In: Easton A, Emery NJ (eds) Cognitive neuroscience of social behaviour. Psychology Press, Hove, pp 115–156

Emery NJ (2006) Cognitive ornithology: the evolution of avian intelligence. Phil Trans R Soc B 361:23–43

Emery NJ, Clayton NS (2001) Effects of experience and social context on prospective caching strategies in scrub jays. Nature 414:443–446

Emery NJ, Clayton NS (2004a) The mentality of crows: convergent evolution of intelligence in corvids and apes. Science 306:1903–1907

Emery NJ, Clayton NS (2004b) Comparing the complex cognitive abilities of birds and primates. In: Rogers LJ, Kaplan G (eds) Comparative vertebrate cognition: are primates superior to non-primates? Kluwer Academic/Plenum Publishers, New York, pp 3–55

Emery NJ, Dally J, Clayton NS (2004) Western scrub-jays (*Aphelocoma californica*) use cognitive strategies to protect their caches from thieving conspecifics. Anim Cogn 7:37–43

Emery NJ, Seed AM, von Bayern AMP, Clayton NS (2007) Cognitive adaptations of social bonding in birds. Phil Trans R Soc B 362:489–505

Gallup GG (1982) Self-awareness and the emergence of mind in primates. Am J Primatol 2:237–248

Gould-Beierle K (2000) A comparison of four corvid species in a working and reference memory task using a radial maze. J Comp Psychol 114:347–356

Hampton RR (1994) Sensitivity to information specifying the line of gaze of humans in sparrows (*Passer domesticus*). Behaviour 130:41–52

Hare B, Tomasello M (2005) Human-like social skills in dogs? Trends Cogn Sci 9:439–444

Hare B, Call J, Agnetta B, Tomasello M (2000) Chimpanzees know what conspecifics do and do not see. Anim Behav 59:771–785

Hare B, Call J, Tomasello M (2001) Do chimpanzees know what conspecifics know? Anim Behav 61:139–151

Hare B, Plyusnina I, Ignacio N, Schepina A, Wrangham R, Trut L (2005) Social cognitive evolution in captive foxes is a correlated by-product of experimental domestication. Curr Biol 15:226–230

Heinrich B (1999) The mind of the raven: investigations and adventures with wolf-birds. Harper Collins, New York

Heinrich B, Pepper JW (1998) Influence of competitors on caching behaviour in the common raven. Anim Behav 56:1083–1090

Heyes CM (1994) Anecdotes, training, trapping and triangulating: do animals attribute mental states? Anim Behav 46:177–188

Heyes CM (1998) Theory of mind in nonhuman primates. Behav Brain Sci 21:101–148

Humphrey NK (1980) Nature's psychologists. In: Josephson B, Ramachandran V (eds) Consciousness and the physical world. Pergamon Press, Oxford, pp 57–75

Jones RB (1980) Reactions of male domestic chicks to two-dimensional eye-like shapes. Anim Behav 28:212–218

Kallander H (1978) Hoarding in the rook (*Corvus frugilegus*). Anser Supp 3:124–128

Kamil A, Maudlin JE (1988) A comparative ecological approach to the study of learning. In: Bolles RC, Beecher MD (eds) Evolution and learning. Lawrence Erlbaum Associates, London, pp 117–134

Kaminski J, Riedel J, Call J, Tomasello M (2005) Domestic goats (*Capra hircus*) follow gaze direction and use social cues in an object choice task. Anim Behav 69:11–18 de Kort SR, Clayton NS (2006) An evolutionary perspective on caching by corvids. Proc Roy Soc 273:417–423

de Kort SR, Tebbich S, Dally JM, Emery NJ, Clayton NS (2006) The comparative cognition of caching. In: Wasserman EA, Zentall TR (eds) Comparative cognition: experimental explanations of animal intelligence. Oxford University Press, New York, pp 602–618

Lefebvre L, Bouchard J (2003) Social learning about food in birds. In: Fragaszy DM, Perry S (eds) The biology of traditions. Cambridge University Press, Cambridge, pp 94–126

Lefebvre L, Nicolakakis N, Boire D (2002) Tools and brains in birds. Behaviour 139: 939–973

Lefebvre L, Whittle P, Lascaris E, Finkelstein A (1997) Feeding innovations and forebrain size in birds. Anim Behav 53:549–560

MacPhail EM (1987) The comparative psychology of intelligence. Behav Brain Sci 10: 645–696

Nichols S, Stich SP (2003) Mindreading: an integrated account of pretence, self-awareness and understanding other minds. Oxford University Press, Oxford

Olson DJ, Kamil AC, Balda RP, Nims PJ (1995) Performance of four seed-caching corvid species in operant tests of non-spatial and spatial memory. J Comp Psychol 109: 173–181

Paz-y-Mino CG, Bond AB, Kamil AC, Balda RP (2004) Pinyon jays use transitive inference to predict social dominance. Nature 430:778–781

Pepperberg IM, McLaughlin MA (1996) Effect of avian-human joint attention on allospecific vocal learning by Grey parrots (*Psittacus erithacus*). J Comp Psych 110:286–297

Pollock B, Prior H, Gunturkun O (2000) Development of object permanence in food-storing magpies (*Pica pica*). J Comp Psychol 114:148–157

Povinelli DJ (2000) Folk physics for apes. Oxford University Press, New York

Povinelli DJ, Eddy TJ (1996) What young chimpanzees know about seeing. Monogr Soc Res Child Dev 61, Serial No. 247

Povinelli DJ, Vonk J (2003) Chimpanzee minds: suspiciously human? Trends Cogn Sci 7:157–160

Povinelli DJ, Vonk J (2006) We don't need a microscope to explore the chimpanzee's mind. In: Hurley S, Nudds M (eds) Rational animals? Oxford University Press, Oxford, pp 385–412

Povinelli DJ, Nelson K, Boysen S (1990) Inferences about guessing and knowing by chimpanzees. J Comp Psychol 104:203–210

Reiner A, Perkel DJ, Bruce LL, Butler AB, Csillag A, Kuenzel W, Medina L, Paxino G, Shimizu T, Striedter G, Wild M, Ball GF, Durand S, Gunturkun O, Lee DW, Mello CV, Powers A, White SA, Hough G, Kubikova L, Smulders TV, Wada K, Dugas-Ford J, Husband S, Yamamoto K, Yu J, Siang C, Jarvis ED (2004) Revised nomenclature for avian telencephalon and some related brainstem nuclei. J Comp Neurol 473:377–414

Scaife M (1976a) The response to eye-like shapes by birds I: the effect of context: a predator and a strange bird. Anim Behav 24:195–199

Scaife M (1976b) The response to eye-like shapes by birds II: the importance of staring, pairedness and shape. Anim Behav 24:200–206

Shettleworth SJ (1998) Cognition, evolution and behavior. Oxford University Press, New York

Smith PK (1996) Language and the evolution of mind-reading. In: Carruthers P, Smith PK (eds) Theories of theories of mind. Cambridge University Press, Cambridge, pp 344–354

Striedter GF (2005) Principles of brain evolution. Sinauer Associates Inc, Sunderland, MA

Tomasello M, Call J (1997) Primate cognition. Oxford University Press, New York

Tomonaga M (2006) Development of chimpanzee social cognition in the first 2 years of life. In: Matsuzawa T, Tomonaga M, Tanaka M (eds) Cognitive development in chimpanzees. Springer-Verlag, Tokyo, pp 182–197

Tschudin AJ-PC (2006) Belief attribution tasks with dolphins: what social minds can reveal about animal rationality. In: Hurley S, Nudds M (eds) Rational animals? Oxford University Press, Oxford, pp 413–436

Vander Wall SB (1990) Food hoarding in animals. University of Chicago Press, Chicago, IL

Watve M, Thakar J, Kale A, Putambekar S, Shaikh I, Vaze K, Jog M, Paranjape S (2002) Bee-eaters (*Merops orientalis*) respond to what a predator can see. Anim Cogn 5:253–259

Whiten A, Byrne RW (1988) Tactical deception in primates. Behav Brain Sci 11:233–277

Zentall TR (2004) Action imitation in birds. Learn Behav 32:15–23

5 Cooperation in Keas: Social and Cognitive Factors

LUDWIG HUBER[1], GYULA K. GAJDON[2], IRA FEDERSPIEL[3], and DAGMAR WERDENICH[2]

1. Introduction

For many decades, researchers have tried to get a glimpse into the "folk physics" of animals, that is, their common sense understanding of how the world works, as well as why it works in the way it does (e.g., Köhler 1927). It has been suggested that animals use a variety of cognitive strategies in understanding such things as space, tools, object categories, quantities, and perhaps causality. Significantly later, this focus of research has been complemented with studies on "folk psychology", that is, what animals understand about the behavior and perhaps mental lives of conspecifics in interactions involving cooperation, competition, communication, and social learning (e.g., Premack and Woodruff 1978). However, in both fields of research, we are far from a common agreement of what animals really understand when they are engaged in adapting or modifying their environment, in their physical or social form, for their own sake. While some researchers believe that many large-brained animals develop an increasing ability to understand causal relationships on accumulating experience (Fujita, this volume; Hauser 2000; Rumbaugh et al. 2000), others warn to over-interpret the animal's apprehension of cause-effect relations underlying the manipulation of animate or inanimate objects (Heyes 1998; Povinelli 2000; Tomasello and Call 1997; Visalberghi and Tomasello 1998).

Visalberghi and Tomasello (1998) have seen a conceptual connection between the understanding of the social and the physical world. Advanced forms of intel-

[1]Department for Neurobiology and Cognition Research, University of Vienna, Althanstrasse 14, A-1090 Vienna, Austria
[2]Department for Neurobiology and Cognition Research, University of Vienna, Althanstrasse 14, A-1090 Vienna, Austria; Konrad Lorenz Institute for Ethology, Austrian Academy of Sciences, Savoyenstrasse 1a, A-1160 Vienna, Austria
[3]Department for Neurobiology and Cognition Research, University of Vienna, Althanstrasse 14, A-1090 Vienna, Austria; Present address: Sub-Department of Animal Behaviour, Department of Zoology, University of Cambridge, Madingley CB23 8AA, UK

ligence may have the potential to find proper solutions in both domains, for instance by abstracting the most relevant features of the problem, that is, by identifying the regularities of and relationships among events. Although the selective pressures leading to such higher forms of intelligence might have been different, the result is the same. In the case of humans the relations they came to understand involved intentionality in the social domain and causality in the physical domain (Tomasello 2000). Understanding intentionality and causality involves understanding the dynamic relation of an antecedent event to a consequent event in terms of some mediating force that somehow explains "why" the first invariably leads to the second. In the physical realm this leads to a much more flexible approach to problem solving as individuals may substitute antecedent events so long as they initiate the same mediating force (e.g., finding a novel way to crack open a nut), and they may prevent consequent events by blocking the antecedent's initiation of the mediating force. In the social realm, understanding mediating forces means understanding that others are intentional beings whose behavior is guided by mediating forces such as goals and perceptions, which also leads to much more flexible problem solving in the social domain.

Of course, causal reasoning as a means of detecting relationships between important events separated in space and time is only one way of "understanding" that is sufficient to adaptively manipulating the environment. A simpler alternative, accessible to many animal species, is associative learning. As a means of linking contiguous events it is a very powerful mechanism, but it is restricted to very short event relationships, in which effects follow causes immediately. If cause and effect are separated in space or time, associative learning looses its functionality. As humans we possess cognitive tools that are able to detect links between important events separated in space and time, thereby assigning cause-effect or means-end relationships to the physical or social world through either observation or insight. Such causal relationships between objects might result in the representation of abstract concepts, like "connectivity" (Hauser 1997), or in the construction of a chain of responses that lead to a goal. To which extent these cognitive tools are spread across the animal kingdom is a challenge to contemporary research.

In recent times, the most popular and stimulating hypothesis is the "social function of intellect hypothesis" put forward to explain the obvious correlation between problem solving capacity and the complexity of the social system in primates (Humphrey 1976; Jolly 1966). Although the original proponents argue that primate cognition arose in response to the demands of a complex social life, some have extended this hypothesis to birds (e.g., Emery and Clayton 2004; Marler 1996). But even in primates might the need to solve "Machiavellian" problems (Byrne and Whiten 1988) in the social world not be the only selective pressure for the evolution of intelligence. The "technical intelligence hypothesis" (Byrne 1997) has been considered as a complementary evolutionary force in terms of technical, mechanical selective pressures for increasing efficiency in foraging behavior. Byrne (1997) argued that increasing efficiency in foraging behavior was critical for great apes, more so than for monkeys, because of their large size, the difficulties of brachiation for distance travel, and their reliance to high quality diet. According to this proposal, solving "Machiavellian" problems caused the haplorhine rise in intelligence and the later hominid one that led to Homo erectus, while the increasing demands for solving foraging problems caused the ape/monkey grade-shift in intelligence (Byrne 1997). Although there is some plausibility in this proposal, especially if one considers the tool use capabilities of chimpanzees and orangutans, there are also some notable exceptions. Capuchin monkeys (Cebus apella) are well known for their nut cracking behavior, not only in captivity, but also in the wild (Visalberghi 1990).

Tool using behavior has been found not only in primates, but also in a great number of other mammals, birds, and even invertebrates, such as Octopus vulgaris (for a review, see Beck 1980). The currently most intensively studied examples are reported from New Caledonian crows (Corvus moneduloides; Weir et al. 2002). But we do not restrict cases of technical or sensorimotor intelligence to cases of tool using and manufacturing, as evidenced by some corvids, among them most obviously ravens (Corvus corax; Heinrich 2000; Heinrich and Bugnyar 2007) and Grey parrots (Psittacus erithacus; Pepperberg 1999). In fact, the ability to understand functional properties of objects, in a range from simple static relationships to complex dynamic ones, and the ability to coordinate actions accordingly, has been demonstrated by animals in other ways of foraging than by tool using.

If it is true that concealed and hard-to-extract food create special problems to animals that lack specialized anatomy to solve them, especially when the need to forage extractively arises seasonally and over a wide range of foods, advanced forms of sensorimotor intelligence would be of great advantage (Parker and Gibson 1977). Therefore, Huber (in press) suggested to extend the "technical intelligence hypothesis" (Byrne 1997; Parker and Gibson 1977) in order to integrate evolutionary scenarios that fit into this framework and are thereby mirroring the one faced by apes.

This view rests on findings from experiments and field observations of keas (*Nestor notabilis*), alpine parrots that are popular for being bold, curious, manipulative and social. As parrots, they rival corvids and primates in relative brain and telencephalic volumes (Iwaniuk et al. 2005). Especially the astonishing combination of ape-like behavioral characteristics caught our attention and motivated us to start a series of rigorous experimental tests of their cognitive abilities in the lab, complemented by field observations under the harsh conditions of their alpine home range.

2. Social and Technical Intelligence in Kea

One of our first studies on cognition in keas revealed the bird's ability to learn from conspecifics how the world works (Huber et al. 2001). Keas were required to open an "artificial fruit", which consisted of a food box with several tricky locking devices that had to be dismantled in order to get access to the encapsulated reward. This kind of task provides an ecologically more relevant mechanical problem than tool-use tasks for most animals because they more frequently engage in extractive foraging techniques without using tools (Whiten et al. 1996). The behavior of keas that watched skilful models that were trained to open the box differed remarkably from the behavior of keas that had no such opportunity to watch a model. Beside some effects of social facilitation and stimulus enhancement, the observers opened five times as many locking devices than the non-observers and one observer opened all three devices within the first two minutes of the test. This cannot be explained by stimulus enhancement because neither did the observers—in comparison to non-observers—prefer the "correct" part e.g. of the split pin (the loop end), nor did they manipulate it longer. Also the comparison of the test performance of the most successful birds in each group showed that the observer's success at the split pin couldn't be explained by focusing at it from the beginning and then responding to it in the most natural way, namely by pulling. Therefore, the birds learned by observation how the locking devices works, e.g. that the split bin can be removed. Non-observers without the knowledge about this affordance of the locking-device were not successful in dismantling the locks (Huber et al. 2001).

The keas capacity to pay attention to the relevant features of a lock was demonstrated in another non-social task where two identical-looking hooks were attached on the front part of a food box. One hook locked the lid of the food box while the other hook was non-functionally fixed beside the lid (Riedesser et al. 2004). During the first trials the kea did not preferred to unhook the functionally device first but quickly learned to do so. When the position of the functional und non-functional look were exchanged by fixing the hooks in a different orientation, all five keas nevertheless unhooked the new functional hook first this time. They did so even when the non-functional lock was fixed entirely on the front of the lid.

The keas also showed superior performance in the string pulling-tasks (Werdenich and Huber 2006) in which birds are required to repeatedly pull up a string hanging on a perch and fix the string with a foot on the perch between pulling actions in order to get access to a reward attached at the end of the string. With the exception of two fledglings, the keas solved the string-pulling task by coordinating bill and a foot in their first trial. The birds appeared to have achieved these effective solutions spontaneously, because the complete act was accomplished in a rapid and straightforward manner, although they had no prior experience in string pulling. In addition, the keas showed high success rates in tasks where they had to discriminate between strings with baited and unbaited objects. In terms of first-choice performance such immediate solutions were not achieved in the crossed strings tasks. But the birds learned quickly after first failure, with only a few errors in the thirty following trials. These results demonstrate the keas competence in means-end use.

In a tube removal-task, our captive subjects also demonstrated their competence to consider spatially well separated items in problem solving (Gajdon et al. 2003): The majority of our subjects managed to remove the tube (rewarded on the inside) from an 70 cm long upright pole within few trials. When we offered two poles with rewarded tubes at the bases and added blue sticks at the end of the poles in a way that either allowed or blocked tube removal, the birds immediately started to lift the tube and to climb the pole with the appropriate pole end for tube removal. However, they did so only after extended exposure to a two-pole apparatus with a small blocking board at the end of one pole, indicating some inference of playful tube lifting and previous experience with the unblocked one-pole apparatus. It might also be that the blocking of tube removal at the board of the first two-pole apparatus started them to construct a more abstract concept of blocking structures and that this allowed them to solve the task with new changes at both pole ends immediately.

The wild keas we tested in Mount Cook National Park, New Zealand, performed less well in the tube removal-task (Gajdon et al. 2004). Despite the bird's persistent manipulation of the tube, only three of more than fifteen birds managed to remove it from the pole, and two of them did so only after demonstration by a trained conspecific (Gajdon et al. 2004). Similarly, most of the wild keas in the park that attempted to open the lids of large rubbish bins failed to succeed, most likely because the did not become aware that they have to bring the lid in a upright position, where it starts to drop backwards (Gajdon et al. 2006) in press). Finally, the proportion of fast string pulling-solvers was smaller than in our captive subjects (Johnston 1999; Werdenich and Huber 2006). Nevertheless, at least some birds succeeded in these demanding tasks.

This demonstrates the outstanding curiosity and manipulative skills of these birds that are considered to be an adaptation to the kea's harsh habitat (Diamond and Bond 1999). Keas are endemic to the South Island of New Zealand where they live in the mountain areas of southern beech (*Nothofagus*) forest, sub alpine scrub and alpine grassland (e.g. Clarke 1970; Jackson 1960, 1963; Wilson 1990). In this highly seasonal environment, the kea became a flexible opportunist (Diamond and Bond 1999) foraging on different parts of one hundred species of plants, invertebrates and carcass. Thereby it extracts food frequently from the ground (plant roots and insect larvae). Indeed, the kea investigates any novel objects it encounters in its habitat showing an almost total lack of neophobia and a rich manipulative repertoire. Its bill is considered to provide a multifunctional, Swiss army knife-like design for scraping, pealing, tearing, probing, levering, pushing, sawing. Such highly explorative and manipulative behavior may increase the chance that subjects will find new food sources. In our case it helps the birds to solve operant tasks. The evolutionary background of this behavior may be similar like that proposed by (Haemig 1989) for increased exploratory behavior of scrub jays from island where food is more likely to become rare and where there is a lower predation pressure than in populations on the mainland. The kea may hold an extreme position in this respect: living on the alpine "island" of an island with virtual no predation pressure before Europeans arrived about 150 years ago.

Like other species that live in environments with rather limited resources, keas show an extended period of dependence from their parents and of adolescence. Keas keep an immature plumage for three years at least while the plumage of kaka—the kea's closest relative in the lowlands were food is more abundant—is not distinguishable from that of adults after half an year of age (both species fledge at about ten weeks of age; Diamond and Bond (1999). And keas do not start to breed before an age of three to four years while kaka do so in their second year of life. Such a harsh environment might rather promote a despotic dominance style than tolerant social relationships. However, long dependence is a major determinant for rich play behavior in bird species (Diamond and Bond 2003) and the kea is one of the few bird species that shows social object play and does so much more than kaka do (Diamond and Bond 2004). This may buffer despotic social style between play partners. Playful object and even social object investigations may increase the likelihood for cooperation not only for reasons of increased probability to detect a cooperative solution. The lack of high foraging pressure in playful situations may also increase the likelihood to overcome runs of cooperative instances with strong payoff asymmetries.[1] Of course this is most pronounced in adolescent individuals that in addition may be even more insensitive for their immediate foraging efficiency due to immature foraging and cognitive skills. These biases for cooperation are overcome in very young individuals that rely on food sources discovered by older birds and that have immature sensorimotor competence: It is not before around eight months of age that keas start to use interrelation between objects such as pulling a string to obtain a reward (Stamm 2006; Werdenich and Huber 2006) or turning over the correct of two covers under which an object was hidden in view of the subject in succession (Pesendorfer et al. 2005).

Due to the steep topography of the kea's habitat, not much is known about the bird's natural foraging and the impact social factors may have on it. But observations at refuse dumps (Diamond and Bond 1991) revealed that juveniles in their second summer of life use efficient social displays such as hunching to obtaining food that older birds discovered while subadults make extensive use of theft as foraging strategy. Adult keas are the most capable scavengers that engaged more in excavating foraging activities than other age classes. This resume provided by Diamond and Bond has to be considered in the light that keas in their first and second year of live gather in flocks while adult males do only sporadically join these groups and thus may not rely to individually preferred play mates to the same degree as do younger birds (own unpublished data). Thus, social foraging may play a pivotal role in juvenile and adolescent birds. It might help naïve birds to find new food sources discovered by flock members. The finder might have to share its food but it might profit from another flock member's future discoveries of new food sources in turn. In some instances such scrounger/producer interac-

[1]A few reports about interspecific feeding by keas indicate the relevance of playful intention in food providing actions. E.g. the keas at our institute were observed to fling items of food out of the aviary when they watched chickens or crows feeding. Similarly a wild kea in Fox Glacier Village was observed to repeatedly flick out meet of a 3.8 m^3 rubbish bin in the direction of a cat (A. Hampton, personal communication 26/11/04. Meat is a preferred food of keas). Such playful repetition of actions that provide some interesting effect is not uncommon in kea.

tions (Barnard and Sibly 1981) might have a playful and highly dynamic social character. For instance, we occasionally observed a flock of juvenile keas digging wholes in a meadow for roots of various species of herbs. The birds frequently rushed to the site where another flock member was digging, displacing the resident and going to dig in the same hole, then running to the site of a next member while another kea reoccupied the vacant site immediately. Thus, the whole flock mutually investigated and enlarged wholes in the meadow and this had an almost although very chaotic character of cooperation.

From this line of argumentation we might conclude that cooperating keas are most likely to be adolescent or young adult birds and that cooperation might occur between playmates or in situations that provide opportunity for playful object manipulation. In sum, the kea displays a fascinating set of dispositions that may accelerate cooperation, even if the birds are not known to cooperate extensively in the wild. It is the birds' flexibility, playfulness and manipulative skills that may provide the basis to acquire relevant experience in order to construct complex behavior (and perhaps cognition) in interaction with its environment (Huber 2006; Huber and Gajdon 2006).

3. Cooperation

Cooperation in animals occurs in a great variety of species and its purpose is explained by ultimate factors of evolutionary aspects (Dugatkin 1997). But when implementing natural selection the evolution of cooperation seems to provide an apparent problem. Why would selection prefer individuals who risk costs to provide benefits to other individuals? It appears that if individuals act cooperatively, they can increase their success and therefore their fitness. From an ecological perspective, stable cooperation can only be achieved if each party's pay-off exceeds those of individual action. The majority of studies deal with these ultimate factors in examining how cooperation has evolved. In this regard four paths have been described to explain the evolution of cooperative behavior in animals: group selection, by-product mutualism, kin selection and reciprocity (Dugatkin 1997; Mesterton-Gibbons and Dugatkin 1992).

Evolutionary explanations have been debated at length, but far less is known about the proximate mechanisms of reciprocity. Especially few attempts have been made to explicitly link work on cooperation to animal cognition. Depending on the specific type of cooperation under focus, and the specific demographics and population structure of the population under investigation, very different cognitive abilities are required for cooperation to be feasible.

Reciprocal altruism—alternating the roles of donor and recipient—has been a well-studied form of cooperation among non-kin because of its intuitive appeal in explaining human cooperation. In human societies, reciprocal altruism is an ubiquitous, integral part of socio-economic behavior, and it is tempting to argue that we have evolved specialized cognitive mechanisms to facilitate its stability, including the systematic detection and punishment of cheaters (Fehr and Gächter

2002). However, despite immense theoretical interest, little empirical evidence substantiates the biological importance of reciprocal altruism in non-human animals, because psychological mechanisms constrain its application in cooperative contexts (Stevens and Hauser 2004; Stevens et al. 2005). Among the cognitive limitations that make reciprocity difficult for animals are numerical quantification, time estimation, delayed gratification, detection and punishment of cheaters, analysis and recall of reputation, and inhibitory control. Stevens and Hauser (2004) argue that most instances of animal cooperation can be attributed to either selfish or indirect benefits via mutualism and helping kin. In cases where it occurs in the laboratory, it is unclear whether the patterns observed would generalize to more natural and less controlled situations.

According to the Prisoner's Dilemma scenario employed by behavioral ecologists to formalize the evolution of reciprocity (Axelrod 1984; Axelrod and Hamilton 1981), mutual cooperation results in moderate reward, but mutual defection leads to low payoffs for both partners. When one cooperates and the other defects, the defector receives the largest possible reward. This implies that mutual cooperation is better than mutual defection, but for an individual partner, there is a sizable temptation to defect.

In situations in which individuals gain immediate benefits by cooperating, cooperation is "selfish". Selfish cooperation or byproduct mutualism is quite common in animal societies. In such cases, partner A benefits from what partner B is doing but would also be doing in the absence of partner A (van Schaik and Kappeler 2006). For byproduct mutualism, neither memory nor individual recognition is necessary (Dugatkin and Alifieri 2002). A recent study demonstrated that cooperation in small groups can evolve when the individual takes the last encounter into consideration regardless of whether the partner is the same or a different one (Pfeiffer et al. 2005).

Those students of cooperation interested more in the proximate mechanisms and in the process of cooperation rather than on the outcome and its ultimate causes, prefer a broad, operationally formulated definition, such as the one by Brosnan and de Waal (2002): *"the voluntary acting together of two or more individuals that brings about, or could potentially bring about, an end situation that benefits one, both, or all of them in a way that could not have brought about individually"* (p. 130). This behavior-oriented definition has the appeal to include the cooperative effort, even if the actors fail to achieve the cooperative goal. On the other hand, it excludes interactions that appear cooperative from the point of view of the outcome, but lack the *voluntary* nature of cooperation (coercion; see below).

Hauser and colleagues (2003) designed a series of experiments in which cotton-top tamarins (*Saguinus oedipus*) could altruistically pull a tool to give food to unrelated recipients without getting food for themselves. Tamarins pulled the food most often for partners that always pulled and infrequently for partners that never pulled. The tamarins, however, cooperated less than 50% of the time, and as each game progressed, the amount of food given dropped. Thus they didn't demonstrate robust reciprocity, as didn't blue jays in a similar situation (Stephens et al. 2002). These birds were tested in game theoretical situations of cooperation. While they did not perform a Prisoner's Dilemma matrix in the first study, they followed this tactic in the latter study with slightly modified conditions. Here the blue jays maintained cooperation due to temporal discounting.

Critical analysis of cooperation in primates reveals that the achievement of cooperation by pursuing a common goal may not only be the result of extensive cognitive capabilities but also a consequence of shared behavior that depends on social constraints. Several studies have demonstrated the importance of social factors on cooperative success. Differences in tolerance, as well as in dominance hierarchy and level of agonistic interactions could explain why coactions in moving heavy stones occurred frequently in Tonkean macaques but was rare in rhesus macaques (Petit et al. 1992). In Tonkean macaques dominance is relaxed, conflicts are generally bidirectional, individuals may easily interact with others regardless of social status, and cofeeding may involve several individuals without much contest (Thierry 1985, 1990; Thierry et al. 1989).

Very high levels of social tolerance and even active food sharing are reported from callithrichid monkeys. Werdenich and Huber (2002) tested common marmosets in a cooperative pulling task and found successful cooperation of dyads being dependent of the specific distribution of roles and differences in social tolerance of higher-ranking individuals. Although all individuals were willing to cooperate with at least one partner, only half of the dyads solved the cooperation task. Examination of the factors that correlated with success in this phase revealed that primarily those dyads cooperated in which the dominant subject took the role of the recipient (scrounger) and the subordinate took the role of the donor (producer). However, in these successful dyads the dominant animal did not force (coerce) the subordinate partner to pull the handle. Rather, the partners of cooperative dyads shared the reward and pulled equally often in both the dyadic training and the cooperation test.

The importance of social factors in problem solving tasks has also been investigated in chimpanzees (Hare 2001; Hare and Tomasello 2004; Melis et al. 2006). Chimpanzees do perform elaborated cooperative behavior in the wild (Boesch 1994; Mitani et al. 2000; Watts 2002) but their cooperative problem solving abilities in captivity have so far not fulfilled the expectations as one would expect from such a highly intelligent species. Presuming that chimpanzee's socio-ecology is more oriented towards competition (gaining access to food or mates) than to maintain cooperative skills, Hare (2001) postulated the "competition cognition hypothesis". It assumes that chimpanzees will demonstrate their greatest skills or motivation in competition rather than in cooperation cognition tasks (Hare 2001). This thesis was strengthened by findings on problem solving abilities, including object choice tests. Chimpanzees performed better when tested under competition than under cooperative situations (Hare and Tomasello 2004). In another study chimpanzee were tested in dyads in a cooperative problem solving task and levels of tolerance (food sharing) were controlled for. As predicted, pairs with high levels of social tolerance spontaneously cooperated while other dyads with low rates of inter-individual tolerance never succeeded (Melis-Pérez et al. 2006).

4. Cooperation in Keas: The Seesaw Paradigm

In our experiments with keas we have been interested in proximate mechanisms of cooperation, that is, the immediate behavioral goals that animals pursue and what they understand about them, which ultimately contribute to the animal's fitness. In these studies we addressed the question whether the keas are cooperating solely on the basis of byproduct mutualism, with reciprocation based on pre-existing features of the relationship between dyads, or on attitudinal reciprocity, with reciprocation based on general social predisposition, or on calculated reciprocity, with reciprocation based on a behavioral one-on-one basis (de Waal 2000; de Waal and Luttrell 1988; de Waal and Brosnan 2006). Besides the cognitive mechanisms we were interested in the social influence on cooperative problem solving. Unlike in most other studies on cooperation the keas were not isolated from each other but observed in their social group. Isolated models restrict the tremendous variety of social interactions (Schuster et al. 1993) that have an important impact on the failure or the success of cooperation. Under natural circumstances cooperation can only have evolved without such restrictions on the behavior of the species and only if all factors that contribute to cooperative behavior are analyzed.

Only one previous study has addressed these questions with keas in an experimental set-up, both at the functional and the mechanistic level (Tebbich et al. 1996). A group of seven captive keas was tested on their ability to coordinate their behavior in order to obtain a food reward and, if so, whether this coordinated behavior results in food sharing, reciprocity or in the asymmetric distribution of rewards among birds. The authors selected an instrumental task that forced participants to react to each other's behavior. A "seesaw apparatus" had to be manipulated by at least two birds performing different roles simultaneously. One bird had to push down a lever while the other one could acquire food from a box (see Fig. 1, for a later version of the apparatus). Therefore, in a single interaction only one bird initially received a reward. Keas were tested in the group and in dyads with changing partners.

The authors reported an astonishingly special and complex form of "cooperation", coercion. In the dyadic tests, three dominant keas aggressively manipulated their respective subordinate partners to open the apparatus. The dominance hierarchy enabled them to maintain the preferred position at the food box and to monopolize the food reward, while subordinate birds were allowed only to manipulate the handle of the apparatus. In the group tests, cooperation emerged from reciprocity in subsequent encounters, because all birds involved in manipulations of the seesaw gained access to food rewards with at least one (subordinate) partner. This was possible because of the non-linear hierarchy of the group, in which each bird had at least one subordinate partner. The higher-ranking individual always obtained the reward and each bird changed its role according to its dominance status. This resulted in a symmetric distribution of the reward, which may explain why participants continued to operate the lever throughout the experiments, even without being forced to do so by others.

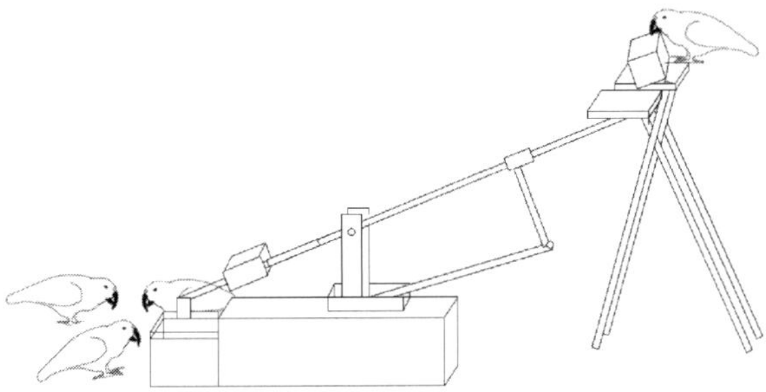

FIG. 1. The brass seesaw apparatus used for the cooperation tests in keas (Federspiel et al. 2005). The basic apparatus was originally designed by Tebbich et al. (1996). It consisted of a lever placed on a wooden box. One end of the lever was a handle or a platform mounted in a way that enabled the keas to sit on it. A transparent plastic lid was attached to its other end closing a food box. One bird had to sit on the handle or pull the lever down with its bill or leg in order to open the lid and enable another bird to obtain food. Once the lever was released, the lid closed again. In a second version devised by Federspiel et al. (2005), the handle was replaced by a platform onto which a wooden cube could be pushed from a stool in order to operate the lever. In this graph a bird is sitting on the stool manipulating the cube, three others are waiting at the food box.

As the authors emphasized, it remained an open question whether this situation with symmetrical pay-offs to all participants would have been stable over a long period. On the one hand, when the youngest individual will loose its ability to displace the most dominant in the group after becoming older, and without any reward would become reluctant to operate the lever of the apparatus voluntarily. In the resulting linear hierarchy, cooperation through subsequent encounters would fade out from the lowest rank upwards.

On the other hand, it remains to be tested if the dominant birds that enforced cooperation on weaker individuals by means of aggressive behavior were based on an understanding of the role of the partner and the mechanism of the seesaw apparatus. Although Tebbich and colleagues (1996) found a complex form of compulsion involving manipulation of the partner to perform a certain type of behavior that benefits only the other, such behavior is also conceivable as a by-product of agonistic group behavior rather than an intentional manipulation of conspecifics. A crucial question here is if the dominant animal foresees a sequence of events by understanding the partner's role and the causal structure involved (Chalmeau and Gallo 1995; Chalmeau et al. 1997; Mendres and de Waal 2000; Visalberghi et al. 2000), or only becomes more aggressive towards the partner when reward is missing. Before this study, the cognitively higher form of forced cooperation on subordinates by means of social manipulation was found in humans only (Alexander 1974).

In order to bring more light into the kea's understanding involved in cooperation or coercion, Federspiel et al. (2005) used the seesaw-cube paradigm for further tests with captive keas in an instrumental cooperation task without any pre-training (Brosnan and de Waal 2002; Petit et al. 1992). A group of ten captive keas (five fledglings, one juvenile, one subadult, three adults) was confronted with a slightly modified brass seesaw (Fig. 1). These keas were kept in the same aviary as those of Tebbich et al. (1996), but none of the keas tested in the study by Tebbich et al. (1996) were used again.

During 23 daily test sessions the keas made 1235 opening attempts, from which in only 81 cases (0.07%) at least one bird gained profit.

At the time the experiments were conducted, there was a linear rank order within the group. The distribution of roles clearly depended on this rank order. In all cooperative interactions at least one of the profiting birds was dominant over the bird opening the seesaw. The highest-ranking bird within the group was able to gain 87% of the overall profit, the lowest ranking only 1% ($\chi^2 = 92{,}3$, df = 2, $p \leq 0.001$). Furthermore, the two highest-ranking birds never performed any opening. As was also found in group tests of chimpanzees (Chalmeau and Gallo 1996), dominant birds maintained the preferred position at the food box and monopolized the reward. Subordinate birds were only allowed to manipulate the lever of the seesaw and to open the food box. However, due to the linear dominance hierarchy, keas failed to demonstrate reciprocity in subsequent encounters, as was the case in Tebbich et al. (1996). Instead, some reciprocity was found by food sharing, because the bird operating the lever was able to profit himself in 56% of the cases by licking butter off of the twigs the profiting bird left behind.

To investigate whether cooperation would also persist if the subordinate birds that operated the seesaw would be prevented from getting some food themselves, thus having only costs without benefits (Brosnan and de Waal 2002; Trivers 1971; de Waal and Harcourt 1992), the reward was changed from buttered twigs to butter pellets in a second group test. As expected, the cooperation faded immediately. The subordinate birds became increasingly reluctant to open the seesaw, and finally stopped to participate at all.

To further investigate the influence of the involved costs and benefits, the seesaw was modified in a third phase. A platform was mounted on the handle, and a stool with a cube on it was placed next to it (Fig. 1). The birds could now operate the lever by either sitting on the platform ("perch solution") or by pushing the cube onto it ("cube solution"). In contrast to the cooperative solution without personal gain, this technical solution offered the possibility to operate the seesaw and to profit from one's own action because the lid remained open after the cube was pushed onto the platform.

Interestingly, Federspiel (2006) found a strong dependence of age in the choice of methods. While the older birds preferred the cube solution from the first trial on (Mann Whitney-U = 3077, Z = −5.862, $p \leq 0.001$), the younger ones preferred the perch solution (Mann Whitney-U = 4200, Z = −2.021, $p \leq 0.043$). This indicates that older keas react sensitively to their own payoff and show great flexibility in applying new methods. The profit was again distributed according to the rank order. However, while older birds switched to a more profitable solution, fledglings maintained their unprofitable behavior, applying the perch solution.

The most obvious explanation is that kea fledglings are insufficiently prepared for the cube technique. In the wild, fledglings compensate lack of technical understanding by showing a variety of social strategies (Diamond and Bond 1991). They rarely discover food on their own and are routinely fed by adults of their group, especially if they display begging behavior. Furthermore, the fledglings showed a greater tolerance towards each other than towards older birds. These egalitarian relationships enabled them to profit from one another. In contrast to the older birds, they could therefore open the seesaw by sitting on the platform and profit afterwards, by licking off butter of the others' bills.

But what was the underlying mechanism in the adults' behavior at the seesaw? Did they understand the affordances of the cube? First of all, there was a significant difference between the numbers of aggressive approaches at the different parts of the seesaw. Ninety-nine of the 215 observed displacements during the Phases 1 and 2 took place within one body length at the seesaw (84% of which at the food box), whereas only 1% happened at the log, from where the birds could reach to the lever ($\chi^2 = 410.4$; df = 3, $p \leq 0.001$). Similar results were found for the third phase: 98% of the 395 observed displacements took place at the seesaw, whereas only 2% happened at the stool, from where the cube could be pushed onto the platform or the lever could be operated ($\chi^2 = 116.2$; df = 1, $p \leq 0.001$). Thus, the subdominant bird being at the log or the stool seemed to have been appeared as being at a "profitable" position to the dominant. For the lower ranking birds these areas were "safe places" where they wouldn't be displaced from. Presumably, the dominant birds expected to profit if the lower ranking animals would stay in these areas.

Another point is supportive for the assumption that the keas have acquired some understanding of their roles. They never opened the seesaw by operating the lever when they were alone. What remained to be tested was their understanding of the mechanics of the seesaw apparatus. To this end, experiments with keas in isolation were conducted. Every bird was confronted with exactly the same setup as during the group experiments, that is, the seesaw and the stool with the cube on top of it. Whereas some birds didn't even approach the seesaw, the two participating birds seemed to consider the direction when pushing the cube off the stool. From four possible directions they selected the correct one significantly most frequently; the two birds pushed the cube onto the platform of the seesaw in 65% and 72% of the cases, respectively ($\chi^2 = 18.8$; df = 3, $p \leq 0.001$; $\chi^2 = 93.0$; df = 3, $p \leq 0.001$).

5. Testing Keas with A Double-sided Box

In order to make a more general statement about the abilities of keas to cooperate in an instrumental task, Dagmar Werdenich (unpubl. data) devised a different setup and tested another sample of captive keas in Vienna. A group of five keas, housed at the Schönbrunn zoo in Vienna, participated in the experiment. The group consisted of one adult female, one adult male, two subadult females and one juvenile male (Table 1). The apparatus was available to the whole group and at least two individuals had to work together to obtain food rewards by performing different roles simultaneously. Only one participant could receive the food, but they had the opportunity to share food with their partners or reciprocally alter the roles. The apparatus consisted of a wire mesh box containing two food bowls that were baited with butter (Fig. 2).

The keas solved this task instantaneously. All five individuals performed the required action within minutes, by quickly adapting the force of pulling to move the bowl to the opposite side. In the following sessions the individuals also learned to coordinate their actions. Over the course of 15 sessions they showed 431 cooperative interactions resulting in more than 900 rewards.

In contrast to the group of keas tested by Tebbich et al. (1996), but similar to the one tested by Federspiel (2006), the dominance hierarchy of the group in the zoo was linear. All five individuals participated in the experiment, with four performing role reversal by taking the role of the food-producer as well as the

TABLE 1. Percentage of produced and scrounged rewards for five keas listed according to their rank position in the group.

Individual	Rank	Sex	Age class	Produced rewards	Scrounged rewards
Aroha	1	F	Adult	43.9%	0.8%
Berliner	2	M	Adult	11.0%	56.1%
Grisu	3	M	Juvenile	16.7%	43.0%
Kari	4	F	Subadult	28.0%	0.2%
Juma	5	F	Subadult	0.4%	0.0%

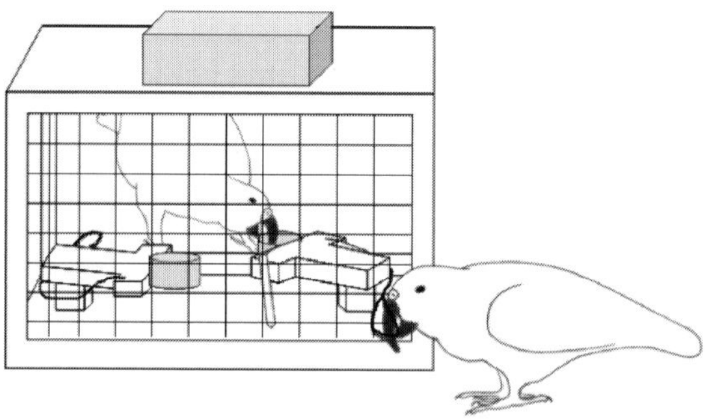

FIG. 2. The apparatus used for cooperation tests in the Schönbrunn zoo in Vienna. Two food bowls were each attached to a lever and a string was connected to each lever. The birds had to pull the string in order to move a lever with a food bowl to the opposite side of the apparatus. A rubber band brought the bowl back to the original position as soon as the pulling individual let go of the string. In this graph, one bird has pulled at the string, which caused the food bowl to move to the opposite side of the box. The kea, who waited behind, is feeding from the food bowl.

role of the recipient. Perhaps due to her low rank, only the lowest-ranking female made accessible food to the others without obtaining any rewards in return. Interestingly, the dominant individuals did not monopolize the apparatus but participated in producing rewards, so that even two lower-ranking birds were able to obtain some rewards. This might have caused the persistence of cooperative interactions for the whole duration of the experiment, which lasted for 15 sessions.

Rewards were never shared and on no occasion did more than one bird obtain food from the bowl. The distribution of reward followed an interesting interaction of sex and dominance (Table 1). Aroha, the dominant individual of the group, was responsible for the largest amount of pulling actions, followed by Kari, the female at the fourth rank position in the group. However, these two females could profit rarely. In contrast, the two males Berliner and Grisu, at the second and third rank position, respectively, obtained nearly all of the food, but produced much less rewards than the two females Aroha and Kari.

Why does cooperation became at least temporarily stable despite the fact that some individuals work for little or no exchange of favors? It is possible that the females persisted for some time in pulling because these activities were rewarding in themselves or because they failed to understand their role in the cooperation task. Since keas are extremely curious and prone to manipulate objects, some degree of manipulation of any object is to be expected regardless of the situation or of what the animals understand of the task. However, this cannot explain why the females didn't stop cooperating for more than a month. In similar experiments with chimpanzees, a young female was also found to cooperate with the dominant male who monopolized the rewards. This female was not forced to cooperate, but her help was recruited and appeared voluntary. As the authors argued, the food reward was not likely a factor for this infant female, because she was not yet dependent on solid foods and was probably rewarded with play (Chalmeau 1994; Chalmeau and Gallo 1996).

The number of pulling actions from which another bird could profit increased after the first session in relation to the number of pulling actions in isolation. While this cooperation rate was only 5.6% in the first session, keas rapidly improved their behavior thereafter (43.7% in the second session), although the absolute number of pulling actions remained equally high. During further 13 sessions the cooperation rate remained at approximately this level.

The keas also learned to take the spatial position of their partner into consideration. In group tests this is especially important because of the dynamic nature of participation with sometimes rapidly changing partners. Nevertheless keas could follow the actions caused by one specific partner and based their behavior on each other. They pulled significantly less when they were alone at the apparatus compared to when at least one kea joined them at the apparatus (Wilcoxon test: $T = 117$, $N = 15$, $p < 0.01$). Note that the keas could move freely in the big aviary, where they had available many objects for behavioral enrichment reasons. Moreover, they adjusted their pulling activity to the spatial position of the participant at the apparatus by pulling significantly more frequently when the partner was sufficiently near the position from where it could reach the food bowl than anywhere else at the box (Wilcoxon test: T = 110, N = 15, $p < 0.01$). This result is especially interesting, because capuchin monkeys successfully cooperated in a pulling task without understanding their partner's role (Chalmeau et al. 1997). In contrast to a later study (Mendres and de Waal 2000), the monkeys did neither put their partner's spatial position nor their behavior into account.

Besides the cognitive mechanisms required for successful cooperation, the social interactions before, during and after the cooperation activities are important. Unfortunately, in many cooperation tests, the partners are tested only in dyads, physically separated from each other (e.g., de Waal and Brosnan 2006). These restrictions not only exclude any group effects but also, even worse, prevent the "cooperating" partners from any direct interactions. The keas did show many direct interactions, among them many agonistic ones. Most aggressive approaches were observed at the apparatus, where dominant individuals displaced subordinate ones from the most attractive parts of the test apparatus. However, and in contrast to the findings by Tebbich et al. (1996), dominant birds did not approach the subordinate partner out of reaching distance from the box or did force them to pull the string. Instead they themselves participated in producing food for others, and together with the two females contributed in maintaining cooperation for the whole duration of the experiment.

6. Summary and Conclusions

From our work with keas it has become obvious that the collaborative solution of instrumental tasks depends on two main factors, the cognitive competence and favorable social conditions. To cooperate effectively, individuals must know what needs to be done and be willing to do it. Until very recently, such full-blown ability to cooperate was seldom found. The recent work of Melis et al. (2006) and Warneken and Tomasello (2006) with chimpanzees may change the picture.

Experimental efforts to induce nonhuman animals to work together in joint tasks have met with mixed success. On the one hand it is not clear whether collaborative failures occurred because animals didn't understand how to solve the tasks or because they were inhibited by the presence of competitors who monopolized the apparatus and appropriated rewards (Silk 2006). On the other hand, collaborative success does not necessarily require understanding of the partner's role or the causal structure of the apparatus. Nevertheless, as the presented findings suggest, keas do perform quite sophisticated forms of coworking. Depending on the instrumental task and the social structure of the group, dominant individuals force subordinate ones to cooperate (Tebbich et al. 1996), switch flexibly between technical and social solutions (Federspiel 2006), or make food accessible to others if they are in the proper position to profit (Werdenich, unpubl. data). Although further experiments are necessary to render our understanding of these forms of coworking in keas more conclusive, these findings fit into tests of their social and technical understanding (Huber 2006; Huber and Gajdon 2006).

The understanding of cooperation in keas will also profit from efforts to investigate the social structure, life history and ecological conditions of this parrot. From our own results, and from what is known from observation of wild keas, it is very implausible that they are driven by altruistic motivations. Rather, the keas are pursuing own goals, by either trying to increase their foraging success or, when satiated, by investigating new objects or seeking opportunities to play. Foraging behavior, particularly of highly cognitive animals such as corvids, primates and cetaceans, is frequently a social activity and as such has received much theoretical interest in the form of the producer-scrounger game (Barnard and Sibly 1981). It is commonplace that in any given population there are individuals differing in their investment in searching for food and in the exploitation of the food found by others. According to the producer-scrounger model, dominant individuals are expected to gain much of the finder's share by either retaining their prey or by stealing the prey of others. However, as we have shown in marmosets, tolerant species may show more complex patterns of role distributions (Werdenich and Huber 2002).

In keas, another important factor influencing the producer-scrounger game is sex. While adult males are more likely to be scroungers if enough producers are around, females may invest as producers for other profits than food. Perhaps different goods and services are exchanged within a network of individuals, as predicted by biological market theory (Noë and Hammerstein 1995). Moreover, in some instances the same currency will have a different relative value to different individuals, based on such factors as the animal's rank, size, or age (Boyd 1992; Seyfarth and Cheney 1988). We may therefore tentatively conclude, that cooperation in keas has evolved as byproduct mutualism, as it was predicted to occur when animals live in "harsh" environments, in which there is an immediate cost or penalty for not acting cooperatively (Mesterton-Gibbons and Dugatkin 1992). Related phenomena of group coordination are group augmentation, where animals directly profit from being in a group (Kokko et al. 2001) and cooperative hunting by lions stalking large predators that could not be taken by a single hunter (Scheel and Packer 1991).

In order to explain the coordination of behavior in foraging groups we might recur to theories of cooperation, producer-scrounger behavior and biological markets. To date, these theories are not well integrated. Our experiments with keas suggest mutual profits to emerge from the collaborative efforts of their proponents to merging the theories.

References

Alexander RD (1974) The evolution of social behavior. Ann Rev Ecol Syst 5:325–383

Axelrod R (1984) The evolution of cooperation. Science 211:1390–1396

Axelrod R, Hamilton W (1981) The evolution of cooperation. MIT Press, New York

Barnard CJ, Sibly RM (1981) Producer and ccroungers: a general model and its application to captive flocks of house sparrows. Anim Behav 29:543–550

Beck B (1980) Animal tool behaviour: the use and manufacture of tools by animals. Garland, New York

Boesch C (1994) Cooperative hunting in wild chimpanzees. Anim Behav 48:653–667

Boyd R (1992) The evolution of reciprocity when conditions vary. In: Hartcourt AH, de Waals FBM (eds) Coalition and alliances in human and other animals. Oxford University Press, Oxford, pp 493–510

Brosnan SF, de Waal FBM (2002) A proximate perspective on reciprocal altruism. Hum Nat 13:129–152

Byrne RW (1997) The Technical Intelligence hypothesis: an additional evolutionary stimulus to intelligence? In: Whiten A, Byrne RW (eds) Machiavellian intelligence II: extensions and evaluations. Cambridge University Press, Cambridge, pp 289–312

Byrne RW, Whiten A (1988) Machiavellian. Clarendon Press, Intelligence

Chalmeau R (1994) Do chimpanzees cooperate in a learning task? Primates 35:385–392

Chalmeau R, Gallo A (1995) Cooperation in primates: critical analysis of behavioural criteria. Behav Process 35:101–111

Chalmeau R, Gallo A (1996) What chimpanzees (Pan troglodytes) learn in cooperative task. Primates 37:39–47

Chalmeau R, Visalberghi E, Gallo A (1997) Capuchin monkeys (Cebus apella) fail to understand a cooperative task. Anim Behav 54:1215–1225

Clarke CMH (1970) Observations on population movements and food of the kea (Nestor notabilis). Notornis 17:105–114

Diamond J, Bond AB (1991) Social behavior and the ontogeny of foraging in the kea (Nestor notabilis). Ethology 88:128–144

Diamond J, Bond AB (1999) Kea: bird of paradox. The evolution and behavior of a New Zealand parrot. University of California Press, Berkeley, CA

Diamond J, Bond AB (2003) A comparative analysis of social play in birds. Behavior 140:1091–1115

Diamond J, Bond AB (2004) Social play in kaka (Nestor meridionalis) with comparison to kea (Nestor notabilis). Behavior 141:777–798

Dugatkin LA (1997) Cooperation among animals: an evolutionary perspective. Oxford University Press, New York

Dugatkin LA, Alifieri MS (2002). A cognitive approach to the study of animal cooperation. In: Allan C, Bekoff M (eds) The cognitive animal: empirical and theoretical perspectives on animal cognition. MIT Press, Cambridge, MA, pp 413–419

Emery NJ, Clayton NS (2004) The mentality of crows: convergent evolution of intelligence in corvids and apes. Science 306:1903–1907

Federspiel I (2006) Social and cognitive aspects of cooperation in keas (Nestor notabilis). Diploma thesis, University of Vienna

Federspiel I, Gajdon GK, Huber L (2005) Do Keas, *Nestor notabilis*, switch from unprofitable cooperative behaviour to alternative solutions? *XXIX International Ethological Conference (IEC)*, Budapest, Hungary, August 20–27

Fehr E, Gächter S (2002) Altruistic punishment in humans. Nature 415:137–140

Gajdon GK, Fijn N, Huber L (2004) Testing social learning in a wild mountain parrot, the kea (Nestor notabilis). Learn Behav 32:62–71

Gajdon GK, Fijn N, Huber L (2006) Limited spread of innovative behaviour in wild kea, Nestor notabilis. Anim Cogn 9:173–181

Gajdon GK, Ortner MT, Huber L (2003) Do keas have insight in a tube removal task? Poster at: ASAB Summer Meeting Personality: Patterns, Mechanisms and Functions. Why are we all different? Grünau, Austria, July 20–23

Haemig PD (1989) A comparative experimental study of exploratory behavior in Santa Cruz Island and mainland California Scrub Jays, Aphelocoma coerulescens. Bird Behav 8:32–42

Hare B (2001) Can competitive paradigms increase the validity of social cognitive experiments on primates? Anim Cogn 4:269–280

Hare B, Tomasello M (2004) Chimpanzees are more skilful in competitive than in cooperative cognitive tasks. Anim Behav 68:571–581

Hauser MD (1997) Artifactual kinds and functional design features: what primates understand without language. Cognition 64:285–308

Hauser MD (2000) Wild minds: what animals really think. Henry Holt, New York

Hauser MD, Chen MK, Chen F, Chuang E (2003) Give onto others: genetically unrelated cotton-top tamarins preferentially give food to those who altruistically give food back. Proc R Soc Lond B 270:2363–2370

Heinrich B (2000) Testing insight in ravens. In: Heyes C, Huber L (eds) The evolution of cognition, MIT Press, Cambridge, MA, pp 289–305

Heinrich B, Bugnyar T (2007) Just how smart are ravens? Sci Amer 296:64–71

Heyes CM (1998) Theory of mind in nonhuman primates. Behav Brain Sci 21:101–148

Huber L (2007). Emulation learning: the integration of technical and social cognition. In: Dautenhahn K, Nehaniv CL (eds) Imitation and social learning in robots, humans and animals: behavioural, social and communicative dimensions. Cambridge University Press, Cambridge, pp 427–439

Huber L, Gajdon GK (2006) Technical intelligence in animals: the kea model. Anim Cogn 9:295–305

Huber L, Rechberger S, Taborsky M (2001) Social learning affects object exploration and manipulation in keas, Nestor notabilis. Anim Behav 62:945–954

Humphrey NK (1976). The social function of intellect. In: Bateson PPG, Hinde RA (eds) Growing points in ethology. Cambridge University Press, Cambridge, pp 303–317

Iwaniuk AN, Dean KM, Nelson JE (2005) Interspecific allometry of the brain and brain regions in parrots (psittaciformes): comparisons with other birds and primates. Brain Behav Evol 65:40–59

Jackson JR (1960) Keas at arthur's pass. Notornis 9:39–58

Jackson JR (1963) The nesting of keas. Notornis 10:319–326

Johnston R (1999) The kea (Nestor notabilis): a New Zealand problem or problem solver? Master thesis. Christchurch, New Zealand, University of Canterbury

Jolly A (1966) Lemur social behavior and primate intelligence. Science 153:501–506

Köhler W (1927) The mentality of apes, second edition. Vintage Books, New York

Kokko H, Johnstone RA, Clutton-Brock TH (2001) The evolution of cooperative breeding through group augmentation. Proc R Soc Lond B 268:187–196

Marler P (1996). Social cognition: are primates smarter than birds? In: Nolan VJ, Ketterson ED (eds) Current ornithology, vol. 13. Plenum Press, New York, pp 1–32

Melis AP, Hare B, Tomasello M (2006) Chimpanzees recruit the best collaborators. Science 311:1297–1300

Melis AP, Hare B, Tomasello M (2006) Engineering cooperation in chimpanzees: tolerance constraints on cooperation. Anim Behav 72:275–286

Mendres KA, de Waal FBM (2000) Capuchins do cooperate: the advantage of an intuitive task. Anim Behav 60:523–529

Mesterton-Gibbons M, Dugatkin LA (1992) Cooperation among unrelated individuals: evolutionary factors. Q Rev Biol 67:267–281

Mitani JC, Merriwether DA, Zhang C (2000) Male affiliation, cooperation and kinship in wild chimpanzees. Anim Behav 59:885–893

Noë R, Hammerstein P (1995) Biological markets. TREE 39:78–90

Parker ST, Gibson KR (1977) Object manipulation, tool use and sensory motor intelligence as feeding adaptations in Cebus monkeys and great apes. J Hum Evol 6: 623–641

Pepperberg IM (1999) The Alex Studies: cognitive and communicative abilities of Grey parrots. Harvard University Press, Cambridge, MA

Pesendorfer MB, Gajdon GK, Huber L (2005) Object permanence in kea (Nestor notabilis). Poster at the XXIX International Ethological Conference at Budapest, Hungary, August 20–27

Petit O, Desportes C, Thierry B (1992) Differential probability of "coproduction" in two species of macaque. Ethology 90:107–120

Pfeiffer T, Rutte C, Killingback T, Taborsky M, Bonhoeffer S (2005) Evolution of cooperation by generalized reciprocity. Proc R Soc Lond B 272:1115–1120

Povinelli DJ (2000) Folk physics for apes. The chimpanzee's theory of how the world works. Oxford University Press, Oxford New York

Premack D, Woodruff G (1978) Does the chimpanzee have a theory of mind? Behav Brain Sci 4:515–526

Riedesser S, Federspiel I, Gajdon GK, Rechberger S, Huber L (2004) The logic of lock in kea. Poster at: Altenberg workshop in animal logic, Altenberg, Austria

Rumbaugh DM, Beran MJ, Hillix WA (2000). Cause-effect reasoning in humans and animals. In: Heyes C, Huber L (eds) The Evolution of cognition. MIT Press, Cambridge, MA, pp 221–238

van Schaik CP, Kappeler PM (eds) (2006) Cooperation in primates and humans. Mechanisms and evolution. Springer, Heidelberg

Scheel D, Packer C (1991) Group hunting behaviour of lions: a search for cooperation. Anim Behav 41:697–710

Schuster R, Berger BD, Swanson HH (1993) Cooperative social coordination and aggression. II. Effects of sex and housing among three strains of intact laboratory rats differing in aggressiveness. Q J Exp Psych 46:367–390

Seyfarth RM, Cheney DL (1988) Empirical tests of reciprocity theory: problems in assessment. Ethol Sociobiol 9:181–187

Silk JB (2006) Who are more helpful, humans or chimpanzees? Science 311:1248–1249

Stamm P (2006) Ontogenese der Exploration und Neophilie beim Kea (Nestor notabilis) (Ontogeny of exploration behaviour and neophilia in kea). Diplomia thesis, University of Vienna

Stephens DW, McLinn CM, Stevens JR (2002) Discounting and reciprocity in an Iterated Prisoner's Dilemma. Science 298:2216–2218

Stevens JR, Hauser M (2004) Why be nice? Psychological constraints on the evolution of cooperation. Trends Cogn Sci 8:60–65

Stevens JR, Cushman FA, Hauser MD (2005) Evolving the psychological mechanisms for cooperation. Ann Rev Ecol Evol Syst 36:499–518

Tebbich S, Taborsky M, Winkler H (1996) Social manipulation causes cooperation in keas. Anim Behav 52:1–10

Thierry B (1985) Patterns of agonistic interactions in three species of macaque (*Macaca mulatta, M. fasciularis, M. tonkeana*). *Aggr Behav* 11:223–233

Thierry B (1990) Feddback loop between kinship and dominance: the macaque model. *J Theor Biol* 145:511–521

Thierry B, Demaria C, Preuschoft S, Desportes C (1989) Structural convergence between silent bared-teeth display and relaxed open-mouth display in the Tonkean macaques (*Macaca tonkeana*). *Folia Primatol* 52:178–184

Tomasello M (2000) Primate cognition: introduction to the issue. Cogn Sci 24:351–361

Tomasello M, Call J (1997) Primate cognition. Oxford University Press, Oxford New York

Trivers RL (1971) The evolution of reciprocal altruism. Q Rev Biol 46:35–57

Uzgiris IC, Hunt JM (1975) Assessment in infancy: ordinal scales of psychological development. University of Illinois Press, Champaign-Urbana

Visalberghi E (1990) Tool use in Cebus. Fol Primatol 54:146–154

Visalberghi E, Quarantotti BP, Tranchida F (2000) Solving a cooperation task without taking into account the partner's behaviour: the case of capuchin monkeys (Cebus apella). J Comp Psychol 114:297–301

Visalberghi E, Tomasello M (1998) Primate causal understanding in the physical and psychological domains. Behav Proc 42:189–203

de Waal FBM (2000) Attitudinal reciprocity in food sharing among brwon capuchin monkeys. Anim Behav 60:253–361

de Waal FBM, Brosnan SF (2006) Simple and complex reciprocity in primates. In: Kappeler PM, van Schaik CP (eds) Cooperation in primates and humans. Mechanisms and evolution. Springer, Heidelberg, pp 85–106

de Waal FBM, Harcourt AH (1992) Coalitions and alliances: a history of ethological research. In: de Waal FMB, Harcourt AH (eds) Coalitions and alliances in humans and other animals. Oxford University Press, Oxford New York, pp 1–22

de Waal FBM, Luttrell LM (1985) The formal hierarchy of rhesus macaques: an investigation of the bared-teeth display. Am J Primatol 9:73–85

Warneken F, Tomasello M (2006) Altruistic helping in human infants and young chimpanzees. Science 311:1301–1303

Watts D (2002) Reciprocity and interchange in the social relationships of wild male chimpanzees. Behavior 138:299–327

Weir AAS, Chappel J, Kacelnik A (2002) Shaping of hooks in New Caledonian crow. Science 297:981

Werdenich D, Huber L (2002) Social factors determine cooperation in marmosets. Anim Behav 64:771–781

Werdenich D, Huber L (2006) A case of quick problem solving in birds: string-pulling in keas (*Nestor notabilis*). Anim Behav 71:855–863

Whiten A, Custance D, Gomez J-C, Teixidor P, Bard KA (1996) Imitative learning of artificial fruit processing in children (Homo sapiens) and chimpanzees (Pan troglodytes). J Comp Psychol 110:3–14

Wilson KJ (1990) Kea: creature of curiosity. Forest Bird 21:20–26

Part II
Ontogeny of Social Cognition

6
Differences Between Acting as if One Is Experiencing Pain and Acting as if One Is Pretending to Have Pain Among Actors at Three Expertise Levels

Hanae Ando[1] and Masuo Koyasu[2]

Abstract. This study aims to examine how acting skills develop by comparing actors at three expertise levels. Actors played four scenes: "not having any pain," "suffering pain," "pretending to feel pain," and "pretending not to feel pain." Their performances were videotaped, and then rated by 46 (Study 1) and another 40 (Study 2) university students. Study 1 revealed that the less experience actors have, the easier it was for the participants to identify the specific scene they were playing. Study 2 indicated that less experienced actors might be so conscious of the audience that their performances are exaggerated, whereas the subtleties involved in the high-quality acting of experienced actors may make their intentions less clear to the audience. The findings suggest that it is important to act in accordance with what is needed in the scene, taking not only the audience but also the setting as a whole into account.

1. Introduction

Actors can communicate their intentions to others more accurately than can we ordinary people without acting experience. Almost everyone would agree to this statement beyond a doubt. Then, how do actors acquire such skills to communicate accurately? How does their acting change as they gain their experience? In this chapter, we investigated the expertise of actors in communicating their intentions by comparing three groups of actors at different expertise levels through two experiments. In conducting these studies, we took up "pain" as a subject of their acting.

[1] Post Doctoral Fellow, Japan Society for the Promotion of Science, 6 Ichibancho, Chiyoda-ku, Tokyo 102-8471, Japan
[2] Graduate School of Education, Kyoto University, Yoshida-Honmachi, Sakyo-ku, Kyoto 606-8501, Japan

1.1. Communication and Acting about Pain

Pain is a subjective sensation that can be difficult to communicate to others. However, accurate communication about pain is necessary to obtain appropriate help from others such as doctors. Although a number of rating scales and questionnaires have been developed that patients can use to describe their pain (e.g., Melzack, 1975; Bieri et al., 1990), there remains the possibility that patients might over- or underreport their pain; they can tell doctors that they are suffering from pain even when they are not, or conversely, may deny experiencing pain even when they are suffering a great deal.

Because it is possible to exaggerate about or understate pain, we use not only verbal information but also nonverbal information when judging the pain of others. Some previous studies have shown that nonverbal behaviors, especially facial expressions, can show pain more accurately than words do (e.g. Craig, 1992; Jacox, 1980; Poole & Craig, 1992). Poole and Craig (1992), for example, found that people estimate others' pain as being less intense when watching facial expressions of those pretending to have pain than when watching facial expressions that truly indicate pain. Effectively pretending to suffer from pain is not necessarily easy, and others might be able to discover our pretense if they can read our facial expressions, especially if they are experienced healthcare providers (e.g., nurses; see Jacox, 1980). Given the complexities involved in pain communication in the clinical setting, it can be said that a high level of acting skill would be needed to communicate pain by actors. That is why we considered "pain" as a subject of acting to find differences between the acting of experienced actors and that of less experienced actors.

1.2. Two Types of Observers in Professional Acting

Professional acting is different from the acting people do in daily life. Two kinds of observers, the audience and the co-actors on stage, are present for actors' acting, while the observers of acting in daily life are only the communication partners. Actors never give a glance at the audience most of the time and behave as if there is no audience; however, their performances are always witnessed by audience, and, in fact, they are very sensitive to the attention of the audience (Brockbank, 1985). Because of this difference, actors are sometimes required to act as a person who is acting.

In some acting situations, the character played by an actor might not be satisfied with what a co-actor is doing, but may need to pretend that he or she is. In such a scene, the actor has to act as a person who is not happy, but is pretending to be so. In other words, the actor in this complex scene needs to convey to the co-actor that he or she is happy and satisfied with the co-actor, yet simultaneously convey to the audience that he or she is only pretending to be happy and satisfied. In such a scene, the message to the audience and the message to the co-actor are different. Performing a double message such as this likely requires a great deal of acting skill.

In conducting studies, we use two types of scenes, simpler scenes that have a single message for the actor to communicate and more complex scenes that ask the actor to communicate double messages. The purpose of this study is to examine the process of acting skill development by comparing actors at different experience levels in their ability to present more straightforward (with a single message) versus more complex (with double messages) experiences around the issue of pain. Specifically, in two studies, we sought to compare three groups of actors (*novice actors* with less than 1 year acting experience, *intermediate actors* with 1 to 5 years experience, and *junior expert actors* with more than 5 years of experience) with respect to their ability to communicate direct experience (pain versus no pain) and more complex experience (pain versus no pain, but pretending otherwise).

In terms of classifying the subject actors, we followed a previous study that investigated the expertise of actors (Ando, 2002). It has been revealed in various fields that learners need strict training for at least 10 years to be experts (Ericsson, 1996), and Noice and Noice (1997) have suggested that this rule is adopted for actors. For this reason, we call a group of actors who have more than 5 years experience "junior experts," not "experts."

2. Videotaping Actors' Performances

2.1. Actors

We videotaped 36 Japanese actors' performances. They were divided into three groups in accordance with length of their acting experience; 12 novice actors with less than 1 year experience, 12 intermediate actors with 1 to 5 years experience, and 12 junior expert actors with more than 5 years experience. In each group, half the actors were male and the other half were female. Mean age, mean period of acting experience, mean time of acting, and mean time of directing of each group are shown in Table 1.

2.2. Scenario

In this study, we used two types of scenes, scenes that have a single message, and more complex scenes that have double messages. Each actor played all four scenes. In all scenes, the main character that actors played are asked by his/her

TABLE 1. Details about actors

Group	Mean age (years)	Mean period of acting experience	Mean time of acting	Mean time of directing
Novice	19.0	6 months	1.1	0
Intermediate	21.8	2 years and 11 months	7.8	1.4
Junior expert	30.3	11 years and 3 months	36.6	9.8

sister whether he/she has a stomachache. In Scene 1 (the "not having any pain" scene), he/she answers "I don't have any pain" because he/she actually does not feel any pain; in Scene 2 (the "suffering pain" scene), he/she answers "I have some pain" because he/she can actually feel pain; in Scene 3 (the "pretending to feel pain" scene), he/she answers "I have some pain" even though he/she is not experiencing any pain; in Scene 4 (the "pretending not to feel pain" scene), he/she answers "I don't have any pain" even though he/she does. Scene 1 and Scene 2 have a single message, and Scene 3 and Scene 4 have double messages. The specific text for these scenes (translated from Japanese) was as follows.

The character you're going to play had a bad stomachache yesterday. It's a holiday today, and the character and his/her sister have planned to go to the movies today. When he/she gets up in the morning, the sister asks him/her "Do you still have a stomachache?"

Scene 1: The character does not have a stomachache now, so please say "I don't have any pain" as the character. When you act, please keep in mind that you have to convey to the audience that you really do not have a stomachache.

Scene 2: The character still has a stomachache now, so please say "I have some pain" as the character. When you act, please keep in mind that you have to convey to the audience that you really have a stomachache.

Scene 3: The character does not have a stomachache now, but he/she is unwilling to go to the movies, so please say "I have some pain" as the character. When you act, please keep in mind that you have to convey to the audience that you are pretending to have a stomachache even though you do not have a stomachache in fact.

Scene 4: The character still has a stomachache now, but he/she does not want to depress the sister, so please say "I don't have any pain" as the character. When you act, please keep in mind that you have to convey to the audience that you are pretending not to have pain even though you do have a stomachache.

2.3. Procedure

Actors participated individually. At first, they read the scenario of one of the four scenes, and were given explanation about the scene. They then practiced the scene for 1 minute. After the practice, they acted in front of a video camera, and their performances were videotaped. The experimenter, the first author, said the line of the sister ("Do you still have a stomachache?"), and the actors acted as if the sister was standing at the place of the video camera. They were instructed to act while sitting still, without using their arms or hands, and their heads and chests were videotaped. After the performances, they were asked to talk about their acting strategies.

We repeated these procedures [(1) 1 minute's practice, (2) performance, and (3) talk about acting strategies] three times for each scene. After three versions of the each scene were videotaped, actors watched the performances on a monitor and chose the single performance that they thought was best. We used only these best performances in Study 1 and Study 2. We repeated this procedure for each of the four scenes, and the order of the scenes was counterbalanced among the actors.

3. Study 1

We conducted Study 1 to verify two hypotheses about the expertise of actors playing scenes that have a single message or double messages. The first hypothesis is about the differences among the three groups of actors. Ando and Koyasu (2004) compared facial expressions of actors with those of non-actors and found that actors could convey their intentions to the audience more strongly than non-actors. Based on these findings, we predicted that acting experience would be associated with ability to convey intention to the audience. That is, the audience would more easily identify the specific scenes when watching performances by the junior expert actors compared to the less experienced actors, and scene identification would be easier when the scenes were performed by intermediate actors compared to novice actors. The second hypothesis concerned the differences among the four scenes. We predicted that scenes that have double messages (Scenes 3 and 4) would be more difficult to perform well than scenes that have a single message (Scenes 1 and 2). Therefore, audience could easily understand the actors' intention when watching scenes with a single message but find it more difficult to understand scenes containing double messages.

3.1. Method

3.1.1. Participants

Forty-six undergraduate and graduate students of a university in Japan participated in this study; 20 were male and 26 were female. Their mean age was 21.4 years old.

3.1.2. Material

We used 144 videotaped performances, which were judged by the 36 actors to be their "best" performances of the four scenes. The videotaped performances were edited so as to start at the moment when the experimenter finished saying the sister's line, "Do you still have a stomachache?" and end at the moment 1 second after the actor finished saying the character's line, "I have some pain" or "I don't have any pain."

The performances of Scenes 1 and 4 (in both scenes, actors said, "I don't have any pain") were mixed, and those of Scenes 2 and 3 (in both scenes, actors said, "I have some pain") were mixed. Therefore, the 72 scenes in which actors said, "I don't have any pain" and the 72 scenes in which actors said "I have some pain" were presented separately to the participants.

3.1.3. Procedure

The experiment was conducted in small groups from 2 to 15 participants. They watched the performances projected on a 100-inch screen and responded to four questions about each performance. When they watched a performance in which an actor said, "I don't have any pain" (Scene 1 or 4), they were asked to indicate whether the actor really did not have any pain (Scene 1) or the actor was pretending not to have pain (Scene 4). Likewise, when participants watched a performance in which an actor said, "I have some pain" (Scene 2 or 3), they indicated whether the actor really did have some pain (Scene 2) or the actor was just pretending (Scene 3). Participants then evaluated: (1) how confident they were in their choices; (2) how real the actor's performance was; and (3) how severe was the pain the actor was communicating. The latter three items were evaluated according to a seven-point scale.

About half the participants watched Scenes 1 and 4 first, and the other half watched Scenes 2 and 3 first. The order of the performances was counterbalanced.

3.2. Results

3.2.1. Scene Identification

Participants were requested to identify the four scenes for every performance of every actor. The average number of correct answers was calculated for each scene and each group of actors (Fig. 1).

Statistical analyses showed that there were significant differences among the three groups of actors. However, the differences found were contrary to the first hypothesis. Specifically in Scenes 3 and 4, differences among the three groups of actors were outstanding. In Scene 3, participants found it easier to identify the specific scene for the novice actors than the intermediate and junior expert actors,

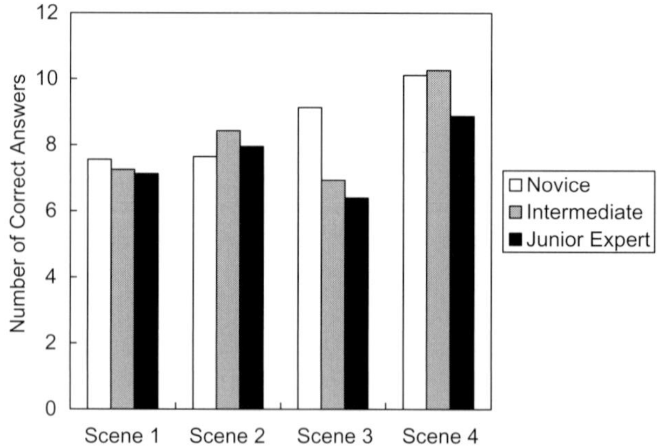

FIG. 1. Average number of correct answers (maximum = 12) participants could identify for each scene

and in Scene 4, they found it easier to identify the specific scene for the novice and intermediate actors than the junior expert actors. In both scenes, it was easiest to identify the specific scene for the novice actors and the most difficult to identify the specific scene for the junior expert actors.

There were also differences among the four scenes that were in contrast to the second hypothesis. All three groups of actors were the most successful in conveying their intention in Scene 4 of all four scenes, which contained double messages. Especially, novice actors were more successful in conveying their intentions in Scenes 3 and 4, which contained double messages, than Scenes 1 and 2, which contained a single message.

3.2.2 Confidence

Participants evaluated how confident they were in their choice of scene identification. The average of their evaluation scores was calculated for each scene and each group of actors, and is shown in Fig. 2.

To summarize the results suggested by statistical analyses, differences among the three actors' groups were found only in Scene 3. For this scene, participants could judge novice and intermediate actors' performances with more confidence than they could junior expert actors' performances. In addition, when identifying performances of novice and intermediate actors, participants could judge scenes that contained double messages (Scenes 3 and 4) with more confidence than the scenes that had a single message (Scenes 1 and 2).

3.2.3. Reality of Performances

Participants evaluated how real the actors' performances were. The average of their evaluation scores was calculated for each scene and each group of actors (Fig. 3).

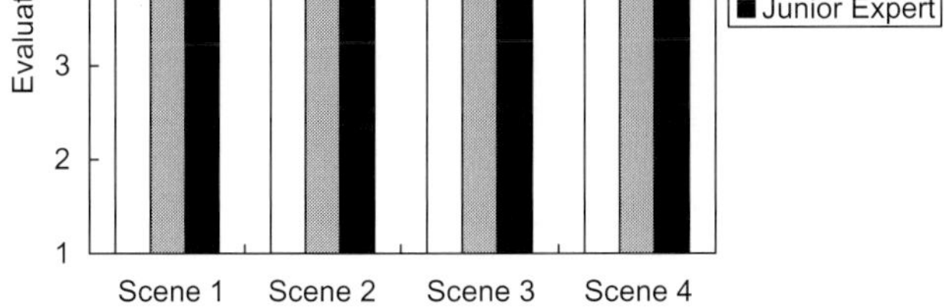

FIG. 2. Average evaluation scores by which participants evaluated how confident they were in scene identification

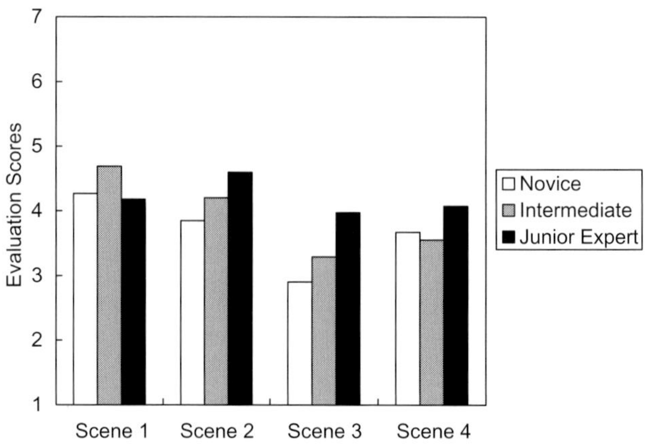

FIG. 3. Average evaluation scores by which participants evaluated how realistic the actors' performances were

Statistical analyses revealed that the performances of junior expert actors were judged as the most realistic, whereas performances of novice actors were judged as the least realistic in Scenes 2, 3, and 4, although in Scene 1, performances of intermediate actors were judged as more realistic than those of novice and junior expert actors. In Scenes 2 and 3, performances of junior expert actors were judged as more realistic than those of intermediate actors, which were judged as more realistic than those of novice actors. In Scene 4, performances of novice and intermediate actors were judged equally, as less realistic than those of junior expert actors.

In terms of differences among the four scenes, we found that novice and intermediate actors' performances in the scenes that contained double messages (Scenes 3 and 4) were less realistic than their performances in the scenes that had a single message (Scenes 1 and 2); however, junior expert actors did not exhibit such a tendency. In sum, performances of novice and intermediate actors became less realistic in the scenes that had double messages compared with the scenes that had a single message whereas junior expert actors did not show such a tendency. That is the reason why junior expert actors excelled in reality when performing scenes that contained double messages.

3.2.4. Severity of Pain

Participants evaluated how severe the pain was that the actors were expressing. The average of their evaluation scores was calculated for each scene and each group of actors (Fig. 4).

For the differences among the three groups of actors, we found that the main effect of the groups of actors was not significant as a result of analysis of variance. For the differences among the four scenes, the results were consistent among three groups of actors: all groups of actors were better at expressing the severity of pain in Scene 2 and most poorly in Scene 1.

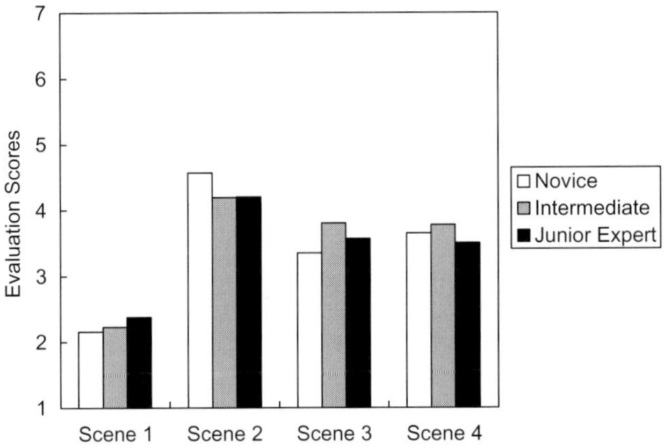

FIG. 4. Average evaluation scores by which participants evaluated how severe was the pain expressed by the actors

TABLE 2. Correlations between evaluation items in study 1

	Scene identification	Confidence	Reality of performance	Severity of pain
Scene identification	1			
Confidence	0.48**	1		
Reality of performances	–0.20*	–0.60**	1	
Severity of pain	0.05	0.22*	–0.18*	1

*$P < 0.05$, **$P < 0.01$

3.2.5. Correlation between Evaluation Items

Table 2 shows the correlation coefficients between the following evaluation items: scene identification, confidence, reality of performances, and severity of pain.

Whether participants could discriminate the scene correctly or not showed a strong association with their confidence in scene discrimination, and was negatively associated with reality of the performances. There was a strong negative correlation between confidence and reality of performances, and a negative correlation between reality of performances and severity of pain. Confidence was significantly correlated with severity of pain.

3.3. Discussion

With regard to the communication of pain severity, actors of all three experience levels communicated the severest pain in Scene 2 and the least severe pain in Scene 1. This result means that none of the groups of actors performed inadequately in regard to expressing pain.

3.3.1. Verification of Hypotheses

The first hypothesis of this study was that the more experience actors have, the more able they would be to convey their intention to their audience. However, in contrast to this hypothesis, at least with respect to scene identification, participants were more accurate when identifying the scenes that novice actors were playing and were least accurate when identifying the scenes that junior expert actors were playing in Scenes 3 and 4. Also contrary to our prediction, in Scenes 1 and 2, the performances by junior expert actors were not identified more accurately than those of intermediate and novice actors. These findings suggest that the less experience actors have, the more obvious it is to the audience which scene they are playing, especially in the more complex scenes that have double messages. This finding was replicated in the analyses concerning confidence ratings, with the results suggesting that audience could identify scenes of novice and intermediate actors with more confidence than those of junior expert actors in Scene 3, although in the other three scenes, no significant differences among the three groups of actors were identified. Overall, and inconsistent with the first hypothesis, the results indicate that in the scenes that have double messages, the less experience actors had, the more accurately and the more confidently the audience could understand the actors' intentions. In the scenes with a single message, there were no differences among the three groups of actors. In short, the findings are not consistent with the first hypothesis.

The second hypothesis was that audience could more easily understand actors' intentions when watching scenes of a single message, and that it would be more difficult to understand actors' intentions in scenes with double messages. The study findings were also inconsistent with this hypothesis. Actors in all three groups were the most successful in conveying their intentions to the audience in Scene 4, the scene that contained double messages. For novice actors, participants identified Scenes 3 and 4 with more accuracy than Scenes 1 and 2. Moreover, it was revealed that participants could identify the scene with more confidence in Scenes 3 and 4 than in Scenes 1 and 2 for novice and intermediate actors.

3.3.2. Reality of Performances

Why were the results contrary to the study hypotheses? The key to solving this question may be related to the reality of performances. In Scenes 2, 3, and 4, the performances by junior expert actors were evaluated as the most realistic whereas the performances of novice actors were evaluated as the least so. The reality of the novice and intermediate actors decreased in their performances in Scenes 3 and 4 in comparison with Scenes 1 and 2, whereas junior expert actors did not display such a tendency. It can be said that the more experience actors have, the more realistically they can perform, especially in scenes that include double messages; that is, the less likely they may be to "overact." Consistent with this explanation, the results of correlation analyses between evaluation items revealed that the reality of a performance had strong negative correlations with scene identification and confidence in scene identification. Therefore, the findings indicate that the more realistically actors perform, the less accurately and the less confidently the audience can identify which scene actors are performing. It is plausible that unrealistic performances are exaggerated ones, making it possible for the audience to understand the actors' intentions with ease and confidence.

In the first hypothesis, we proposed that the actors with more experience would be better able to convey their intentions to the audience. Instead, however, we found that the more experience actors have, the more realistically they are able to perform. As a result, junior expert actors could not convey the specific scene they are acting in as well as the novice and intermediate actors, whose performances were less realistic and perhaps more exaggerated. The second hypothesis was based on the idea that it would be more difficult for the audience to understand actors' intentions in the scenes with double messages than in those with a single message. However, we found that the reality of performances by novice and intermediate actors decreased in the scenes with double messages; consequently, the audience could easily understand their intentions in comparison with the scenes with a single message.

3.3.3. Balance between Two Messages

Why, then, did the junior expert actors perform in a realistic way at the expense of conveying their intentions to the audience? It is plausible that the junior expert actors could perform both in a realistic way and in an unrealistic and exaggerated way, but they chose the realistic way because they judged this to be more suitable for the scenes. Some junior expert actors actually performed both in a realistic way and in an unrealistic way, then selected realistic performances as the best ones from the three performances they have finished. An example of what such an actor (*A*) said to the experimenter (*E*) when choosing the best performance is as follows.

E: Which performance was the best?
A: I think the second performance.
E: The second?
A: Yes. The first performance was funny.
E: Was it funny?
A: It was easily understandable.

This actor talked as above when he chose his best performance for Scene 3. He judged the first performance as easily understandable and funny, and he did not consider this understandable performance as the best one. The same actor made the following comments when choosing his best performance for Scene 4.

E: Which was the best performance?
A: The third performance.
E: The third?
A: Yes. The first and the second performances were, what can I say, well, too understandable. Yes, they were. They were funny.
E: You laughed at your performances when watching them, didn't you?
A: Because I understood the intentions too easily.

Thus, for both Scenes 3 and 4, this actor specifically did not select the performances for which he understood the intentions easily as the best performance.

In Scenes 3 and 4, each character had to act in such a way as to deceive the co-actor, his or her sister. If their performances were unrealistic and exaggerated, the audience could easily understand that the characters were pretending, as would the sister character. Performances that very obviously conveyed to the audience that the characters were pretending would also likely to convey to the sister that the actors were pretending; such performances would not be suitable for the situation in which the character wanted to deceive the sister. Less experienced actors might make too much of a message to the audience and make light of a message to the co-actor. Although it would be very difficult to strike a balance between a message to the audience and a message to the co-actor, junior expert actors might manage to do this. To address this issue, we conducted Study 2.

4. Study 2

In playing a scene that features double messages, we assume that an actor must not only convey to the audience that the character he or she is acting is pretending, but also conceal from the co-actor that the character is pretending. This is a very complex task, and we therefore predicted that more experienced actors would be more effective than less experienced actors in striking the appropriate balance between these two communications.

4.1. Method

4.1.1. Participants

Forty undergraduate and graduate students of a university in Japan participated in Study 2. None of the Study 2 participants had participated in Study 1. Eighteen were male, and 22 were female. Their mean age was 20.4 years old.

4.1.2. Material

We used 72 videotaped performances from Study 1, these being the best performances of 36 actors for Scenes 3 and 4. We used only scenes that featured double messages.

4.1.3. Procedure

Unlike Study 1, where participants viewed the performances in a group setting, the Study 2 participants viewed the performances by themselves, because there were differences among the Study 1 participants in the time they needed to evaluate performances. Participants watched the performances on a computer screen, and evaluated each with respect to four qualities: (1) how successful the character was in deceiving his or her sister; (2) how successful the actor was in conveying to the audience that the character was pretending; (3) how real the actor's performance was; and (4) the overall quality of the actor's performance. They evaluated these four domains on seven-point scales. The order of the performances was counterbalanced.

4.2. Results

4.2.1. Success in Deceiving the Sister Character

Participants evaluated how successful the character was in deceiving his or her sister for every performance. The average evaluation scores are presented in Fig. 5.

Statistical analyses revealed that junior expert actors were more successful than intermediate actors, and intermediate actors were more successful than novice actors in deceiving the sister in Scene 3. In Scene 4, evaluation scores for intermediate actors did not differ from those for novice actors, and were lower than those for junior expert actors. As for differences between Scenes 3 and 4, novice actors were more successful in Scene 4 than in Scene 3, whereas intermediate and junior expert actors were more successful in Scene 3 than in Scene 4 in deceiving the sister.

4.2.2. Success in Conveying Pretence to the Audience

Participants evaluated for every performance how successful the actor was in conveying to the audience that the character was pretending. The average evaluation scores are presented in Fig. 6.

Statistical analyses showed the results were in contrast to the results of the success in deceiving the sister character. In Scene 3, novice actors were more successful than intermediate actors, and intermediate actors were more successful

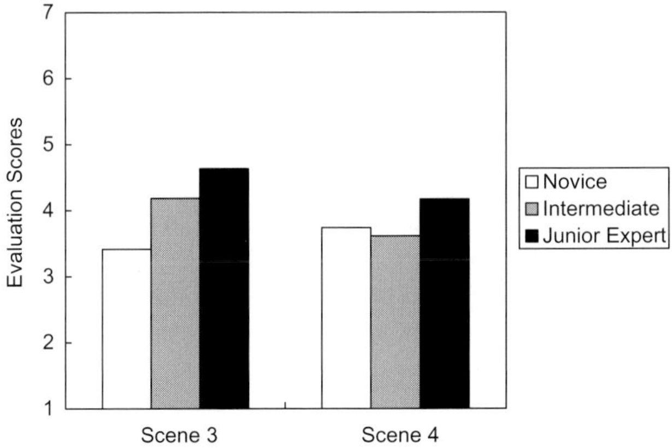

FIG. 5. Average evaluation scores by which participants evaluated how successful the characters were in deceiving their sisters

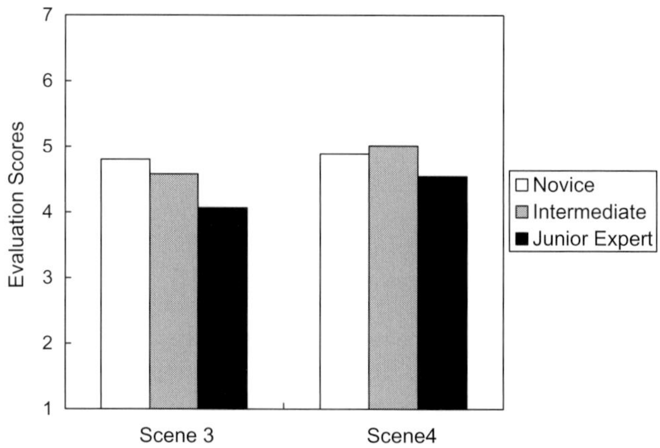

FIG. 6. Average evaluation scores by which participants evaluated how successful the actors were in conveying to the audience that the characters were deceiving their sister

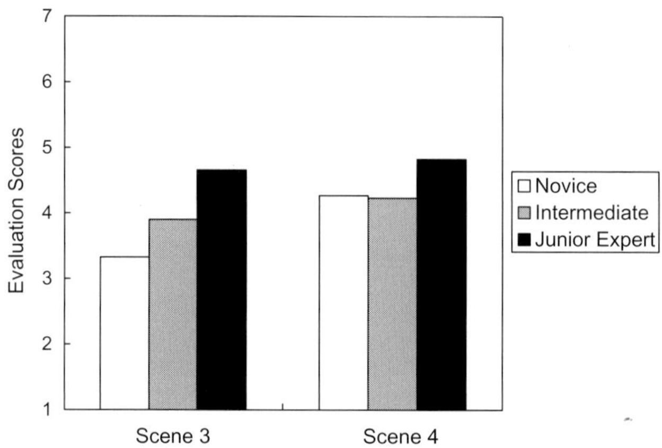

FIG. 7. Average evaluation scores by which participants evaluated how realistic the actors' performances were

than junior expert actors in conveying pretence to the audience. In Scene 4, novice and intermediate actors were evaluated as more successful than junior expert actors. As for the differences between Scenes 3 and 4, intermediate and junior expert actors were more successful in Scene 4 than in Scene 3 in conveying pretence to the audience.

4.2.3. Reality of Performances

The average ratings of reality of actors' performances are presented in Fig. 7.

Statistical analyses revealed that the performances of junior expert actors were evaluated as more realistic than those of intermediate actors, which were evaluated as more realistic than the performances of novice actors in Scene 3. In Scene 4, performances of intermediate actors were as realistic as those of novice actors, and less realistic than those of junior expert actors. As for the differences between Scenes 3 and 4, performances of novice and intermediate actors were evaluated as more realistic in Scene 4 than in Scene 3.

4.2.4. Quality of Performances

The averages of the participants' ratings of the quality of the actors' performances are shown in Fig. 8.

Statistical analyses showed similar results to the results of the reality of performances. In Scene 3, performances of junior expert actors were better than those of intermediate actors, which were better than those of novice actors. In Scene 4, performances of intermediate actors were evaluated as good as those of novice actors, and those of junior expert actors were evaluated as better than those of the less experienced two groups of actors. For the differences between Scenes 3 and 4, novice actors performed better in Scene 4 than in Scene 3, although intermediate and junior expert actors performed Scene 3 as well as Scene 4.

4.2.5. Correlations between Evaluation Items

Table 3 shows the correlation coefficients between the following evaluation items: success in deceiving the sister, success in conveying pretence to the audience, reality of performances, and quality of performances.

Whether actors succeeded in conveying to the audience that the character was pretending has a strong negative association with the other three evaluation items. The other three items showed strong positive correlations with each other.

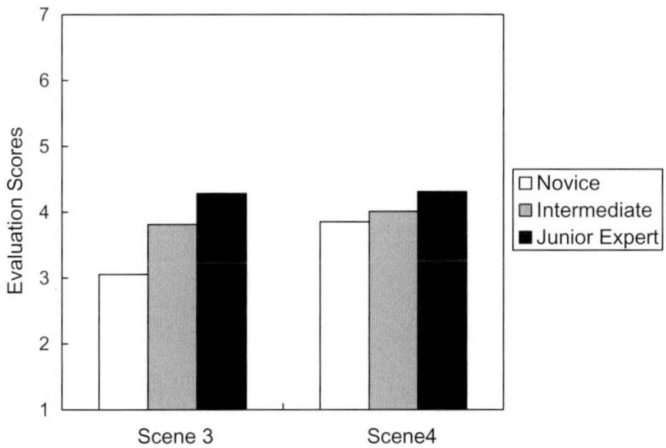

FIG. 8. Average evaluation scores by which participants evaluated how good the actors' performances were

TABLE 3. Correlations between evaluation items in study 2

	Successfulness in deceiving	Successfulness in conveying	Reality of performance	Quality of performance
Successfulness in deceiving	1			
Successfulness in conveying	–0.88**	1		
Reality of performances	0.78**	–0.67**	1	
Quality of performance	0.80**	–0.54**	0.88**	1

$^{**}P < 0.01$

4.3. Discussion

The results of correlations between evaluation items revealed that if the audience could easily understand that the character was pretending to deceive his or her sister, the participants also thought that the sister would also easily notice this intended deception. In Scenes 3 and 4, the actors had to play the role of a character who wanted to deceive his or her sister; therefore, performances in which the sister would easily detect this deception were inadequate for these scenes. For that reason, the success in conveying pretence to the audience might have a strong negative correlation with the reality of performances and goodness of performances. In other words, the more easily the audience can understand that the character is pretending something, the less realistic and the lower the performance is evaluated.

Novice actors might take only the audience into account as a receiver of their message, and might be so conscious of the audience that the audience could easily understand that the characters they were playing were pretending to deceive his or her sister. However, such performances are not suitable for the scenes that contain double messages, and were evaluated as both unrealistic and having less quality. On the other hand, junior expert actors can pay attention to both the audience and the co-actor as receivers of their messages, and can effectively convey to the audience that the character they are playing is trying to deceive his or her sister. That is, they are able to strike a balance between the message to the audience and the message to the co-actor; therefore, their performances were evaluated as more realistic and better than those of the less experienced actors. In the scenes that contained double messages, it is very important to strike a balance between a message to the audience and a message to the co-actor on the stage, something novice actors are less able to manage.

In the results of all four evaluation items, evaluation scores for intermediate actors did not differ from those of novice actors in Scene 4, whereas all three acting groups differed from each other in Scene 3. In terms of quality of performance, intermediate and junior expert actors in Scene 3 were evaluated as good as in Scene 4. On the other hand, novice actors were evaluated more highly in Scene 4 than in Scene 3. It is plausible that novice actors could perform Scene 4 better than Scene 3, as well as intermediate actors performed Scene 4;

therefore, novice and intermediate actors did not differ in all evaluation items in Scene 4.

5. General Discussion

In contrast to the Study 1 hypotheses, novice actors were rated as doing better than junior expert actors in conveying their intentions to the audience in the scenes that had double messages; this might be because junior expert actors considered it unsuitable to reveal that the characters they were acting were pretending. In the scenes with double messages, there are two types of receivers of actors' messages: the audience and the co-actor. Therefore, if actors are very obvious in showing that the character is trying to deceive the co-actor, not only the audience but also the co-actor can discern their deception with ease.

Novice actors were so conscious of the audience that their performances may have been very exaggerated, making it easy for the audience to understand their intentions. However, in this situation, the co-actor would also easily find out what the actors were experiencing. Such exaggerated performances are unsuitable for these scenes with double messages and were evaluated in Study 2 as unrealistic and as being not so good as the performances of the more experienced actors. On the other hand, junior expert actors appeared to strike a balance between a message to the audience and a message to the co-actor, and as a result, the audience was less able to understand their intentions in comparison with the case of novice and intermediate actors, but they could deceive the co-actor better than novice and intermediate actors. Such performances were suitable for the scenes and thus were evaluated as both realistic and good.

Some of the junior expert actors performed in an exaggerated way once in three acting sessions, and it is indeed plausible that they can perform both in a realistic way and an exaggerated way. For example, in a slapstick comedy, junior expert actors would perform in an exaggerated and unrealistic way. Thus, more experienced actors are able to adapt their acting manner according to the needs of the individual scene.

In many areas of human performance, "flexibility" is one of the main characteristics of experts (Feltovich et al. 1997). Expert baseball hitters, for example, can adapt to many different kinds of pitches, thrown from different angles, and at different speeds, whereas novice hitters might not be able to hit curve balls. In addition, some previous studies have revealed that experts consider more information in circumstances to decide their action than novices do. McPerson and Thomas (1989) compared novice and expert players in tennis, and found that expert players decided how to hit the ball considering many things; their own position, positions of their opponents, the level of the opponents, and so on. Along these lines, experienced actors might consider more information than less experienced actors and adapt their acting to the needs of each specific scene. To be a successful actor, it is important to take not only the audience but also the situations on stage into account and change the acting manner according to what is needed in the scene.

Acknowledgments. This research was supported by the Ministry of Education, Science, Sports and Culture, Grant-in-Aid for Exploratory Research obtained by the second author (No. 15650045, 2003–2005). We would like to express our gratitude to Professor Mark P. Jensen, University of Washington, for his helpful comments to the manuscript. We also thank Shinsuke Mouri, Motoko Noguchi, and Mutsumi Kimura, Graduate School of Education, Kyoto University, who assisted with the experiments.

References

Ando H (2002) Engeki no jukutatsuka: kyakuhon no yomitori kara engi-keikaku, engi-suikou made. [The expertise of an actor: script reading, performance planning, and actual performance.] Jpn J Psychol 73:373–379

Ando H, Koyasu M (2004) Engeki keiken no umu ni yoru mikaku hyoujou no hyoushutsu narabini engi no sai. [Differences in facial expressions of tastes between actors and laypersons.] Cognit Studies 11:61–74

Bieri D, Reeve RA, Champion GD, Addicoat L, Ziegler JB (1990) The Faces Pain Scale for the self-assessment of the severity of pain experienced by children: development, initial validation, and preliminary investigation for ratio scale properties. Pain 41: 139–150

Brockbank P (1985) Introduction: Abstracts and brief chronicles. In: Brockbank P (ed) Players of Shakespeare: essays in Shakespearean performance by twelve players with the Royal Shakespeare Company. Cambridge University Press, Cambridge, pp 1–10

Craig KD (1992) The facial expression of pain: better than a thousand words? Am Pain Soc J 1:153–162

Ericsson AK (1996) The acquisition of expert performance: an introduction to some of the issues. In: Ericsson KA (ed) The road to excellence: the acquisition of expert performance in the arts and sciences, sports and games. Erlbaum, Mahwah, NJ, pp 1–50

Feltovich PJ, Spiro RJ, Coulson RL (1997) Issues of expert flexibility in contexts characterized by complexity and change. In: Feltovich PJ, Ford KM, Hoffman RR (eds) Expertise in context: human and machine. MIT Press, Cambridge, MA, pp 125–146

Jacox AK (1980) The assessment of pain. In: Smith WL, Merskey H, Gross SC (eds) Pain: meaning and management. SP Medical & Scientific, New York, pp 75–88

McPherson SL, Thomas JL (1989) Relation of knowledge and performance in boys' tennis: age and expertise. J Exp Child Psychol 13:307–325

Melzack R (1975) The McGill Pain Questionnaire: major properties and sourcing methods. Pain 1:277–299

Noice T, Noice H (1997) The nature of expertise in professional acting: a cognitive view. Erlbaum, Mahwah, NJ

Poole GD, Craig KD (1992) Judgments of genuine, expressed, and faked facial expressions of pain. J Pers Social Psychol 63:797–805

7
Homo Negotiatus: Ontogeny of the Unique Ways Humans Own, Share and Reciprocate

Philippe Rochat[1] and Cláudia Passos Ferreira[2]

Social animals need to share space and resources, whether sexual partners, parents, or food. Sharing is indeed at the core of social life. Humans, however, of all social animals, have distinct ways of sharing. They evolved to become *Homo Negotiatus*; a species that is prone to bargain and to dispute the value of things until some agreement is reached.

In this chapter, we discuss, in the perspective of ontogeny, how children become *Homo Negotiatus*. Our goal is to explore the nature of what makes human ways of sharing unique compared to other animals. For this, we look at how children develop a sense of ownership and a propensity to negotiate with others. This development, we believe, is revealing of the distinct human ways of owning and sharing. Our intuition is that these particular ways of owning and sharing form the fundamental core of what it means to be human. It determines how we grow and how we relate to each other, the origins of our distinct social mind.

There is one main idea driving the chapter. This idea is that ways of sharing and owning are inseparable from particular ways of sensing and knowing the self. They are like two sides of the same coin. From this main idea, we propose a theory postulating that the developmental origins of owning, sharing, and of the sense of self in children are conceptually inseparable. Based on empirical and clinical observations, we speculate and try to specify that owning, sharing and the sense of self develop in parallel. Our goal here is to map the concomitant development of owning, sharing and the self in children.

The chapter is organized as follow. First, we propose and describe different levels of sharing. This distinction articulates the fundamental difference between sharing by coercion and sharing by negotiation that is the trademark evolved by our species. Second, we discuss that to negotiate, as opposed to sharing via brute force and coercion, entails a particular sense of who we are in relation to others. We then present and interpret observations on how infants become *Homo Negotiatus*. Finally, we discuss the parallel emergence of negotiation and theories of mind in children between 3- and 5 years of age.

[1]Department of Psychology, Cognition and Development, Emory University, 318 Psychology Building, 532 Kilgo Circle, Atlanta, GA 30322, USA
[2]Institute of Social Medicine, State University of Rio de Janeiro, Rua São Francisco Xavier, 524, 7° andar, bl. D/E Maracanã, RJ 20550-900, Brazil

In all, here we consider the psychological origins of the human property sense as well as of the particular ways of sharing by negotiation that we view as a major trait that is unique to our species and at the core of human social life.

1. Levels of Social Sharing in Early Ontogeny

All social animals share but they don't do so in the same way. To understand the variety of sharing that pervades social life, it is necessary to distinguish different levels of behavioral organizations that are biological as opposed to psychological in their determination. By biological, we mean behaviors that are automatic and non-intentional. By psychological, we mean behaviors that are intentional and conscious. This distinction is admittedly delicate and elusive but can be made more explicit by considering the levels of behavioral determinants manifested by infants in the course of early development, in particular the first year of life (Rochat 2007).

At birth and during the first 6 weeks of life, infants manifest primarily pre-adapted or "built-in" action systems that allow them to adapt to the circumstances of the environment and to tap into resources they depend on to survive. Neonates and even fetuses during the last trimester of gestation express highly complex sensory-motor organizations. These organizations fulfill basic survival functions such as feeding or the orientation toward particular features in the environment (Reed 1982). For example, at birth infants suck preferentially on certain nipples, they orient to sound and root with mouth open toward tactile stimulations, they are more enticed to track a face with canonical as opposed to scramble features (Morton and Johnson 1991; Rochat 2001; Rochat and Senders 1991). This complex behavioral organization is biologically rather than psychologically determined in the sense that at this level infants are functioning on the basis of pre-determined action systems that are "instinctual" or "obligatory" (pre-reflexive) rather than "contemplative" or "intentional" (see Rochat 2007 for further discussion regarding such distinction). At this level, no representation of goals, nor any expectations regarding what should happen next over time are yet involved.

By 2 months, however, things change and infants manifest more than such instinctual, obligatory, and biologically pre-determined functioning. Infants become less reactive, less stimulus-bound, and more exploratory in their interaction with objects. In relation to people, 2-month-olds begin to show and construct a sense of *shared experience* in face-to-face interaction. They begin to smile back and show the first clear sign of primary inter-subjectivity (Trevarthen 1979). This can be construed as the psychological birth of the infant (Rochat 2001).

From this point on and in relation to people, infants are not simply functioning and reacting. They are actively engaged in the assimilation and alignment of their own subjective experience with the subjective experience of others. They share experiences that are constructed in interaction, typically face-to-face exchanges, with affectively attuned others and in the context of affective resonance (Hobson 2003; Stern 1985).

Parallel to the emergence of experiences constructed in interaction with others, infants also begin to function differently in relation to physical objects. They begin to explore objects. They assimilate objects to their own actions, learning from the perceptual effects they cause by acting on objects (Rochat 2001, 2007). For example, by 2 months, and not prior, infants begin to explore systematically the auditory consequences of their own sucking behavior as they explore a musical pacifier introduced in their mouth (Rochat and Striano 1999b). Their behavioral functioning is not merely responsive or focused on the here and now of perception. Rather, it becomes oriented toward what should happen next, increasingly driven by particular expectations. Likewise, interpersonal exchanges are also increasingly driven by reciprocity principles and social expectations (Rochat and Striano 1999a).

By 7 months, infants show initiatives in trying to influence interpersonal exchanges, by-passing mere passive responding. They become actors and creators in their social transactions. For example, when an adult, in the midst of ongoing proto-conversation suddenly adopts a still face, from two months of age infants show emotional distress and dismay. However, from 7–9 months of age, facing the same circumstances, infants begin to show initiatives in trying to re-engage the still faced person. They lean forward staring at her, call her, pull her cloth or clap hands, clearly with the aim in mind of having the person snap out of her frozen state to re-instate the playful flow of proto-conversation (Rochat and Striano 1999a; Striano and Rochat 1999, 2000).

By this age, infants will also present objects for shared attention. Infants will openly call for attention and frequently check whether others are attentive to what they do with objects. This is particularly evident in all children by 9 months, their relative propensity to engage in joint attention correlated with the developmental emergence of first words and symbolic functioning by the second year (Bruner 1983; Rochat 2001; Rochat and Callaghan 2005; Tomasello 1995; Tomasello and Farrar 1986).

This major development has been extensively documented. It corresponds to the emergence of secondary inter-subjectivity, namely the emergence of referential communication with others about objects in the environment that occurs by the second half of the first year (Bruner 1983; Trevarthen 1979).

By 9 months (9th month revolution or "miracle" according to Tomasello 1995; 1999) infants begin to engage in bouts of joint attention with others as they engage in the exploration of an object. They bring objects to the attention of others and track others' attention in relation to what they do with objects. This new triangulation between the child, another person, and an object of shared attention breaks away from face-to-face exchanges. It makes these exchanges looser and more flexible. Interestingly, it is also associated with a new sense of exclusivity and *possession* first applied to people, then eventually generalized to objects. We have here the putative origins of a property sense expressed by the young child.

By 8–9 months, as infants begin to manifest triadic engagement in reference to objects (i.e., joint attention but also social referencing), they manifest concomitantly a new weariness when encountering strangers, what is described as the 8th

month's anxiety (Spitz 1965). Infants by this age show first evidence of selective attachment and affective bounding to the primary caretaker(s). They also manifest a new fear of separation that is the counterpart of attachment (Bowlby 1969).

2. First Affective Investment into Objects

From this time on (8–9 months), infants begin to invest much affectivity with particular physical objects. Winnicott (1982) provides a complex analysis of the emergence of what he calls "the transitional object", starting at approximately 9 months of age. With the transitional object, whether a blanket, a doll or any other suck-able, hug-able, and transportable physical object, infants suddenly devote particular closeness and a need to cling to them. It is the new expression of a strong affective investment, an affective projection and the binding of affects into a physical object (affective binding). The young child uses such affective projection, in part, to cope with temporary separation from their mother or any primary caretaker. For Winnicott, by the end of the first year the child finds in such objects of devotion a way to cope with separation anxiety, a comforting external entity that becomes companion of their forays away from the secure base of the mother.

At the origins, transitional objects are an affective means, created by children, that allows them to behave with independence and to explore the world outside the primary sphere of fusion with the mother. Literally, it helps them to make this transition away from the mother's secure sphere. They are also probably the affective roots of the sense of material possession.

Following Winnicott's approach, transitional objects are the primitive objects of possession as some kind of a re-incarnation of the mother. The comfort of the mother is transferred and projected into the object that now functions for the child as a substitute to cope with temporary separation. The child's attachment to the mother is transferred to this particular object that becomes transitional.

Transitional objects are, by definition, objects that have value, particularly high *affective value*. They contrast with any other toys or physical things that the child encounters and plays with by the fact that they are affectively invested. The child becomes attached to them as part of themselves. This prefigures the propriety sense that becomes generalized by the end of the second year when the child starts to claim "mine" far beyond their mother.

The infatuation and obsession associated with transitional objects are, to some extent, commensurate to the emotion and affect the child projects onto them. These objects are endowed with new meanings. From being distinct physical toys with particular affordances, they become objects of comfort, endowed with an affective affordance invented by the child. These objects are physically distinct, like any other objects, but become special because the child endows them with the potential to evoke comfort and satisfaction. These objects now have a particular affective value causing a sense of attachment and ownership. We can speculate that this is the origin of the experience of ownership rights over an object, the very beginning of the property sense "proper". From then on, the child can develop a capacity to evaluate, to compare the relative value of objects that are more or less invested affectively. Objects are transformed into "fetish" standing for comfort and security. This new affective meaning attached to the object can sometime persist beyond childhood, into adulthood and through the lifespan of the individual.

Possession therefore implies the projection of affects into the object. By virtue of this affective projection, the object is transformed into an emotional investment that transcends the perceptual experience of its physicality. From physical, the object also becomes affective and this is the psychological bedrock, the first tangible sign of a property sense in the child.

3. From Possession to Negotiation

From the expression of possession and exclusivity, the affective investment onto selective objects of attachment by 9 months, follows a developmental step that is unique to the species. This step emerges by the middle of the second year and corresponds to the progressive inclination children manifest in asserting ownership over things. This opens up the possibility to bring them into sharing space.

Probably the most conspicuous manifestation of such development is in the early use of possessives in language acquisition (Tomasello 1998). By 20 months children become linguistically explicit in their claim of ownership over things. When they relentlessly say "Mine!" by 2 years, they not only mean that it is theirs or that it should be given to them. They also mean that "it is nobody's but mine", in other words that "it is not yours . . .". Such expression is an assertion of power by the child over the object, not just for itself, but in relation to others. "Mine!" is a statement associated with the so-called "terrible two's", a period of defiance and self- assertiveness in the young child who tries to overcome separation anxiety, gain independence as well as social control. But this is also the child's entrance into the adult culture of reciprocal exchanges. It is an expression of exclusivity that actually transcends simple possession. It opens up the possibility for gifts and exchanges since such processes presuppose an explicit and public sense of possession to enable its relinquishing. In other words, it creates the possibility of gift or exchange that is a human trademark. This development entails yet another level of sharing, the level that humans evolved as a species and that each normally constituted child develops to enter the reciprocal culture of his or her parents.

In summary, we proposed various levels of sharing developing in the first 18 months of human life. By two months infants by-pass mere pre-determined functioning to assimilate situations in the environment and generate expectations about what should happen next. This is a first, original step toward "owning" perceptual experience and gaining experiential control over objects and people. By nine months, with the new propensity to share attention and become triadic with others in reference to objects in the world, infants develop also a new infatuation and selective attachment to certain objects, including people. This is the first projection of affect that is the foundation of a property sense. The mechanisms of such projection and the determinants of such development remain under specified and more research is needed. Finally, during the second year, children develop the additional inclination to eventually relinquish what they feel attached to, bringing objects of possession into a space of potential exchange. This, we see as the developmental step that transcends mere possession and marks the child's entrance into the reciprocal culture of his parents. This is when the child becomes *Homo Negotiatus*. Interestingly, this marks also the time when children become less attached and exclusive with a particular object. As Winnicott notes, from approximately 3 years of age, the affective value of the transitional object diffuses and becomes distributed among multiple objects of possession. Affectively invested objects become collective rather than personal and exclusive. They now exist for the child in an interpersonal space of negotiation.

4. Coercion versus Negotiation

Here, we would like to emphasize two fundamentally different categories of sharing. Both entail some sense of ownership or at least rudiments of a property sense, but they are ontologically different because of the psychological and interpersonal processes they entail. However, these two categories of sharing rest on radically different principles. They correspond respectively to sharing by *coercion* and sharing by *negotiation*. We discuss them in turn.

Sharing by coercion pervades nature. It corresponds to a transfer via brute force of what one feels owns or could be owned. It obeys the principle of the lion share: the stronger prevails and gets the most, if not all. With coercion, relative strength, power and assertiveness are the resolving factors of conflict of interests on a particular resource.

Although this kind of sharing is determined by a quantifiable and rather predictable variable (i.e., relative physical strength), it can become complex in instances of bluffing, alliances, and the appeasement of conflicts among individuals (see for example de Waal 1989 in relation to chimpanzees). Many animal species show coalition, the projection of strength via threat, even structural changes in physical appearance (sudden and temporary color change, particular postural displays accentuating physical attributes via hair or tail erection for example) to impress others and influence the sharing of resources while reducing the actual occurrence of physical abuse or fight.

In contrast, sharing by negotiation is unique to humans. It corresponds to a consensual transfer of property among individuals by ways of exchange, one giving and the other receiving. In negotiation, the constraint is not brute force as in coercion. The constraint is to reach some kind of mutual agreement. By definition, negotiation does not abide directly to the principle of the lion share, although perceived power might influence the terms of the agreement reached among sharing protagonists. The weak is more inclined to agree than the stronger, an overwhelmed warrior is typically more eager to settle with his victor than the reverse. However, if strength, force, and power play a major role in any conflict resolution, their role is reduced in negotiation, constrained by a different principle which is *reciprocation*.

Reciprocation is more than the simple tit for tat principle, by which if one gives the other gives back. It entails constant evaluation and tracking of what is exchanged. It also entails agreement and a negotiated sense of fairness that becomes explicit in either the acceptance or the refusal of a bid in the exchange process. Note again that although coercion and negotiation rest on opposite principles, there is a fussy zone between them. A negotiation always has a coercive dimension as the particular strength and background of the negotiators always play a role. Negotiation will be conducted differently depending on the relative strength, reputation or acquired power of the protagonists. However, what makes negotiation particular as a process is the fact that the outcome is agreement, an *inter-subjective agreement on values*.

These values are complex because they do not only pertain to the things exchanged but also to the protagonists of the exchange themselves: whether he or she is relatively tough, understanding, assertive, kind, generous, or on the contrary privy and cheap. The exchange is a public revelation of the person, his or her social inclination, status, and personality. It is the main public arena in which we reveal to each other, the main contributor to the building of reputation among peers, which is the primary concern of humans. To be human is indeed to be concerned about reputation (Rochat 2006, 2008 in press).

Coercion and negotiation both entail a sense of ownership. It entails the sense that something either belongs or could belong to the self; that something can be relinquished, lost, or given; received by the self, taken or given to the self. They both entail a particular sense of self. The property sense and the way possession can be transferred, lost or gained, entails at minimum discrimination between self and world, but more specifically a discrimination between self and others.

One cannot own if one doesn't know who she is, or at least make the difference between herself and others. One owns and claims property necessarily in relation to others. However, sharing by coercion or negotiation each entails a fundamentally different sense of the self. Negotiation implies a sense of self that is continuous over time and a perspective that is situated among other perspectives.

In general, the sense of self that is entailed in the sharing by coercion is temporary, grounded in the immediacy of perception and action. In contrast, the sense of self that is entailed by negotiation is more continuous over time, grounded in memory and the building of long-term reputation. In sharing by negotiation the self gains situation and continuity in relation to others. Also, in the context of negotiation, possession and the claim of ownership have different meaning compared to possession and its claim in the context of coercion. It gives the owner *social power*, the potential to re-enter negotiation, to relinquish what is possessed, the power to barter, eventually even the power to give and show generosity. As shown by early anthropologists like Mauss (1952/1967) or Malinowski (1932), following the pioneer work of Franz Boas on native North American tribes, small scale traditional societies from all over the world are organized around gift systems. In such systems, individuals acquire properties for relinquishing it following particular rituals. By ways of elaborate gifting rituals, individuals build social reputation as well as mutual trust with others that each gift will be reciprocated.

If sharing by coercion or negotiation each entails a fundamentally different sense of the self, it also entails a different construal of others. In sharing by coercion, others are just objects among objects, objects that cling to things. Those endowed with superior force just help themselves whenever they covet something, oblivious of others, helping themselves and always getting the lion share. Essentially, in the coercive process of sharing there is no deep thinking about others and how one relates to other individuals. There are no meta-thoughts, nor any kind of perspective taking involved. What predators experience as agent of the sharing is no more than physical resistance, clinginess, and maybe defiance. But it is a physical exchange, a straightforward causal chain of events made of resistance and overcoming force. No mental state consideration or mind reading is involved. The opposite is true in negotiation.

Sharing by negotiation involves mutual monitoring and mental state consideration. Each protagonist has to track and consider the mental state of the other to decide on the next bid with the ultimate goal to come to an agreement regarding the value of the thing at stake. Emotional expressions are read in reference to desire or beliefs. *Mind reading* is involved, a reading that is mutual, not just surface observation of behavior. Negotiation involves meta-representational abilities that are unique to human, mental reflection leading to propositions such as "he thinks that I think that he feels that we should come to some kind of agreement". It involves the kind of representational self-others reflection that is the mental trait of *Homo Negotiatus*.

So how do children become *Homo Negotiatus*? In the last part of the chapter we account for such development, in particular in relation to change in the early sense of self (self-consciousness) and of others (theories of mind).

5. Becoming Homo Negotiatus and Member of a Self-conscious Species

Negotiation is what happens when we bargain with others, whether ideas, feelings, or objects. Once again, it is the process that captures most exhaustively what human transactions are all about. It is also in this process that human self-consciousness develops, the objectified and conceptual sense of self one has in relation to others, the kind of meta-representation about the self that leads to embarrassment, shame or guilt (Rochat 2008 in press).

By becoming *Homo Negotiatus*, children develop the basic prerequisite of a sense of property as well as self-consciousness. As John Dewey writes: "... the 'Me' cannot exist without the 'Mine'. The self gets solidity and form through an appropriation of things which identifies them with whatever we call myself ... Possession shapes and consolidates the 'I' ..." (Dewey 1922, p. 116).

We argue that human self-consciousness and negotiation are mutual by-products, two expressions of the same development.

In general, what is unique in human transactions is the drive to find agreement with others, to compromise, or not to compromise on all matters, whether affective, intellectual or material. Humans are constantly trying to come to closure with deals, opening the possibilities of new ones. In this process of negotiation, we form knowledge about others as much as we form knowledge about who we are in relation to others. Self-consciousness as the representation of how others perceive and evaluate oneself is a by-product of this process.

Negotiation is the major probing ground by which we weigh ourselves in relation to others. It is also, ultimately, how we figure how much we weigh in the mind of others, how much relative social proximity and how much recognition we have in the eyes of others. The way people respond to our bargain tells us how important we are to them. Inversely, the way we respond and deal with others tells *them* how important they are to us. The point is that negotiation is a permanent game of reciprocal evaluation between self and others. But how does it come about in development? At what point in development do we become *Homo Negotiatus*?

Negotiation in ontogeny finds its roots in the first reciprocal exchanges between infant and caretaker starting in the middle of the second month after term birth. This is indexed by the emergence of socially elicited smiling in proto-conversation with others (so-called primary intersubjectivity). In this new face-to-face communicative context, the child engages in a give and take of affects that implies a turn taking format that is the pragmatic or communicative pre-requisite format of negotiation. In bartering and in proto-conversation alike, one makes a bid and the other takes it or turns it down. The mother smiles, and the child can respond by either a smile or by a frown, he can look toward or look away. There is fundamentally an alternation of bids among the protagonists in the exchange. Furthermore, there is continuity in the exchange as it unfolds, in the same way that there is continuity in bartering and negotiation. A history unfolds, as prior bids determine future bids.

In the affective proto-conversation that emerges unambiguously by 2 months, not prior, we find the primal form of mutual exchange. It is from this alternating and reciprocal frame that infants develop to become *Homo Negotiatus*. The difficult question is then, how and what happens next?

Infants are born *from* and are immersed in *Homo Negotiatus* culture, but they are not born *Homo Negotiatus*. The alternating and reciprocal frame of proto-conversation is encouraged and provided by attuned and responding caretakers (Stern 1985). However, this is not sufficient. To become *Homo Negotiatus*, infants need to develop on their own initiative, pushed by a force that comes from them. We proposed elsewhere (Rochat 2001) that this bootstrapping force in human development might originate from a basic dilemma, a constitutive tension between the propensity to explore and roam about the environment and the urge to maintain proximity with others.

By 9 months, infants are channeled to resolve this basic dilemma by including others in their roaming and exploration of the environment. Infants by this age work hard at incorporating the attention and gaze of others in their foray. They do all they can to captivate others and include their gaze in their exploration. They begin to solicit social attention onto themselves and onto what they are trying to achieve. This is a crucial step in the development of negotiation and a source of budding self-consciousness.

In their attempt at resolving their basic dilemma, infants are eventually *constrained* or channeled to objectify themselves in the gaze of others. They are constrained toward self-objectification as they have to make themselves noticed and to present themselves to others as object of attention and intention. This is indeed the beginning of self-objectification, hence of self-consciousness. Note that this triadic objectification of the self could not occur if others, in particular adults, were not themselves attuned to the attention and intention of the child. Self-objectification can only develop in a *community* of already intentional and self-conscious individuals. Comparative research shows that non-human animals, even close primate relatives do not engage in joint attention and intentional exchanges such as deictic pointing, at least to the levels humans do (Tomasello and Call 1997). This is obviously a pre-requisite condition for the child to become *Homo Negotiatus*.

In this fundamental process of social-attention-getting in order to resolve the constitutive tension between proximity seeking and exploration, infants discover the *social power* that is attached to objects of possession. With the intermediary of objects, infants learn to control the attention of others, capturing this attention toward themselves, the experiential warrant of their social proximity and intimacy.

Children discover that objects are the means by which they can control their sense of social inclusion and recognition, the means by which they ultimately can fulfill their basic affiliation need. They discover that by owning, they can bring what they own in a space of exchange and negotiation. In exchange and negotiation, infants gain further control of others' attention. They also gain further leverage in promoting themselves and gauging their own social worth.

There is clearly a deep incentive to own and claim property as it allows the child to negotiate and accessorily to gain social leverage and control of their own situation in relation to others. By two, children understand explicitly the social power and leverage attached to property, and this is the long-term outcome of early reciprocal exchanges emerging by 2 months. The motivational background of this development is, in general, the basic need to affiliate and maintain proximity with others.

In human development, negotiation is the main process by which we co-construct what we are as persons. This process develops early but gets a new life by the second birthday when children become explicit in *claiming property*. They discover social power in bringing their claim of ownership into negotiation space. Interestingly, it is also at this age that they begin to manifest an explicit *conceptual awareness* regarding who they are, an objectified sense of self as "me" when for example they identify themselves in the mirror (Lewis and Brooks-Gunn 1979; Rochat 2003).

By this age (2–3 years), children also begin to *identify* themselves with others. They are able to consider themselves as differentiated, yet similar to others as in the case of their expression of empathy that is more than simple emotional contagion (Decety and Jackson 2006; Eisenberg 1989; Zahn-Waxler et al. 1992;). All these capacities coalesce by the end of the second year, a time when the child begins to claim property and becomes *Homo Negotiatus* proper. They all correspond to the constitutive elements of negotiation, the basic process by which the self is co-constructed in relation to others.

6. Theories of Mind in Development

If negotiation is a privileged probing ground of what we are in relation to others, it is also a privileged source of knowledge about others, namely the construal of what is on the mind of others, in relation to the self but also in relation to the world at large. Negotiation is a privileged source of so-called "theories of mind" (also called, probably more appropriately, "folk" or "people" psychology). This psychology revolves around the understanding of the thoughts, emotions, beliefs, desires and intentions that underlie other people's actions.

In the heart of negotiation, there is the constant conjecturing and factoring of what is on the mind of others in order to predict and figure their behavior, but also their decisions and valuations in the process. It is also by this constant conjecturing that we probe how others relate to us, always trying to figure our place in the mind of others. In general, in negotiation, one conjectures others not only for what they are as psychological entities endowed with beliefs and wants, but also for what they reflect of one's self-worth.

Much research documents how children come to construe others as having beliefs that can be either the same or on the contrary different from their own; that someone might have a false belief about something the child knows is not true. In the developmental and comparative literature, the ability to construe the false belief of others is considered as the acid test for the existence of theories of mind (Wellman 2002; Wimmer and Perner 1983).

Typically, developmental studies show that it is only by 5 years that the child can figure that someone else has a false belief about the state of thing in the world, beliefs that are different from their own. By 3–4 years, the majority of children do not. At such young age, children have a hard time decoupling and inhibiting their own belief when considering others'. They generalize and assimilate from their own, egocentric perspective.

In a recent study, we confirmed that this developmental transition has a universal character (Callaghan et al. 2005). We found remarkable developmental synchrony between 3 and 5 years in children growing up in 5 highly contrasted cultural contexts: Canada, Samoa, Thailand, India, and Peru. In all cultures, 80% of 3 year-olds failed the classic false belief task as 80% of 5-year-olds passed it. This is a clearly unified developmental trend.

So, between 3 and 5 years, children develop a sophisticated understanding of what is on the mind of others, construing the representations held by others that guide their behavior and determine their world's view: what they hold as being either true or false, desirable or undesirable, realistic or unrealistic. One can assume that when children begin to construe others in this way, going beyond the surface information of their behavior and infer mental states, they also have more sophisticated ways of construing themselves as sentient individuals.

Interestingly, by 3 years children begin to manifest self-conscious emotions, including shame, guilt, pride, or empathy (Kagan 1984; Lewis 1992). This development appears to pre-figure the development of theories of mind applied to others, although both entail sophistication in meta-representation. In a sense, self-conscious (secondary) emotions such as shame or empathy do express meta-representational abilities but that appear first applied in relation to the self. Theories of mind research, in particular the false belief test, suggests that within a few months of developmental time, these meta-representational abilities are generalized to the construal of others. If that is the case, the question is what makes this development possible?

It is likely that theories of mind are actually a spin off of the insatiable drive children have to come to agreement and closure with surrounding others, constantly engaging in emotional trading and bargaining, for better or for worse. Children are constrained to conjugate with others, share resources but also primarily obtain from others in order to survive. This affective as well as material game is set from the outset but changes dramatically in the course of early development.

From the high social dependence of the newborns endowed with pre-adapted action systems (e.g., feeding, orienting systems), infants develop to become more autonomous, yet still highly dependent of their social surrounding. This dependence changes rapidly in forms. Starting at 2 months, we have seen that the format of negotiation begin to be the main engine of children's developing sense of autonomy in relation to others, in other words, the developing sense of themselves as an independent, sentient agent in their social world. Again, negotiation is an emergent property of social exchanges that in humans are based on principles of reciprocity, aside from potentially being also selfish and coercive.

7. Conclusion: Negotiation and Theories of Mind[1]

Negotiation is essentially a conversation that with development is increasingly initiated by the child in the form of bargaining. The child acts to push against and explore the limits of the "*No*" as René Spitz claimed years ago. The toddler runs away toward cliffs, cars, and treacherous places, probing how they will be run after by presumably pressing adults to be picked up and saved. They explore the limits at which others will intervene by either helping or hindering their action. In other words, they act to probe their social world, but more importantly,

[1]This last concluding section as well as the ideas of the preceding two are taken from a book by Rochat (2008 in press), "*Others in Mind—Fear of rejection and the social origins of self-consciousness*".

to probe their situation in this world: how much people care about them and how much intimacy they are capable of generating and controlling in others. This is the main game most evident by the second year but already budding by the second month. It is a game that never leaves us as grown ups.

It is interesting to note that the Latin root of the noun negotiation or the verb to negotiate comes from a contraction of *neg* meaning "not" and *otium* meaning "leisure". Thus, negotiation has the original meaning of the antithesis of leisure, in other words, of time free from the demands of work. This original meaning of the term is rather counterintuitive as we spend most of our time, whether at work or in leisure, questing for agreement and closure with others. This is an endless game that pervades all of our lives. It is as part of this quest that interpersonal values are established, the values of actions and gestures that specify the degree of our affiliation and intimacy with others. Theories of mind take their roots in this process, not the reverse.

Interpersonal needs (intimacy and affiliation) have precedence over the development of theories of mind. These needs determine theories of mind and this is particularly evident when considering the development of active sharing. This development constrains children to construe the mental states of others, to figure their desires, their beliefs and value systems. Children develop such capacity as a necessary requirement for negotiation and active sharing. In fact, children develop theories of mind in the context of learning the rules of constant negotiation and active sharing with others. By engaging in negotiation, children learn as much about themselves, in particular their affective situation in relation to others, as they learn about the mental states of others. The negotiating process channels the child toward the construal of others' mental states, not the reverse.

We collected some data in 3 and 5 year-old children, before and after they succeed in the false belief task, on their ability to negotiate. We observe that children at 3 years of age, not passing the false belief task, show little flexibility and reciprocity understanding in negotiating a barter deal with an adult experimenter.

In our little experiment, the child was given a large collection of small stickers that he could take home if he wished. The Experimenter gave himself a smaller collection of stickers that were much bigger in size and brighter. Both child and experimenter agreed that the experimenter's stickers were much nicer. The Experimenter then asked the child if he or she wanted a sticker of her nicer collection. All children agreed of course and then the Experimenter asked: "what would you give me for one of my sticker?"

Children were invited to barter stickers from their collection. Following the child's offer and according to the experimental procedure, the Experimenter systematically refused any first or second barter deal, eventually accepting it by the third. We were interested to see the extent to which children tended to modify their bid to barter following the refusal by the experimenter. In other words we were interested in the relative flexibility of the child in the negotiation process. What we found is that by 3 years, when children still failed to construe false belief, they also fail to appropriately modify their bartering offering to somehow revive the negotiation with the Experimenter. Typically, 3-year-olds repeatedly offered the sticker that was turned down by the experimenter, demonstrating rigidity or fixedness in their response. In contrast, by 5 years of age, when the large majority of children pass the false belief test, children do demonstrate much more flexibility and appropriate negotiation adjustment by increasing their offer following the Experimenter refusal.

Our observations clearly indicate that the development of negotiation skills parallels, even probably causes the development of theories of mind as measured by the false belief task. Although we don't have supporting data yet, my hunch is that negotiation, as a trademark of the human environment to which children must adapt, forms the facilitating context in which theories of mind come to life.

Negotiation as a reciprocal social adjustment process does call for some construal of others' mental state. Children grow to become *Homo Negotiatus*, and the rest follows, including theories of mind (Rochat 2005). Negotiation precedes and constrains progress in the construal of what others have on their mind, particularly the construal of what they represent about us, the representation of *who we are*. It is the core process by which children can become reciprocating members of a culture that rests on the inter-subjective sense of values, on the agreement and constant bargaining regarding the values of things, whether physical objects, ideas, or affects.

In conclusion, we hope to have made the case that negotiation is at the core of what makes the variety of human cultures human, as opposed to non-human. This is what children develop to acquire the human social mind necessary for their enculturation.

Acknowledgments. Our gratitude to Britt Berg for her diligent help in the editing the manuscript. While writing this chapter, the first author was supported by a 2006 John Simon Guggenheim Memorial Foundation fellowship.

References

Bowlby J (1969) Attachment. BasicBooks, New York

Bruner JS (1983) Child's talk. Norton, New York

Callaghan T, Rochat P, Lillard A, Claux ML, Odden H, Itakura S, Tapanya S, Singh S (2005) Synchrony in the onset of mental-state reasoning: evidence from five cultures. Psychol Sci 16:378–384

Decety J, Jackson PL (2006) A social-neuroscience perspective on empathy. Curr Direct Psychol Sci 15:54–58

Dewey J (1922) Human nature and conduct. An introduction to social psychology. Carlton House, New York

Eisenberg N (1989). The development of prosocial values. In: Eisenberg N, Reykowski J, Staub E (eds) Social and moral values: individual and societal perspectives. Lawrence Erlbaum Associates, Inc, Hillsdale, NJ, pp 87–103

Hobson P (2003) The cradle of thought. Oxford University Press, Oxford

Kagan J (1984) The nature of the child. Basic Books, New York

Lewis M (1992) Shame: the exposed self. Free Press, New York

Lewis M, Brooks-Gunn J (1979) Social cognition and the acquisition self. Plenum Press, New York

Malinowski B (1932) Argonauts of the Western Pacific: an account of native enterprise and adventure in the archipelagoes of Melanesian New Guinea. Routledge & Sons, London

Mauss M (1952/1967) The Gift: forms and functions of exchange in archaic societies. Norton, New York

Morton J, Johnson MH (1991) CONSPEC and CONLERN: a two-process theory of infant face recognition. Psychol Rev 98:164–181

Reed ES (1982) An outline of a theory of action systems. J Motor Behav 14:98–134

Rochat P (2001) The infant's world. The Developing Child Series. Harvard University Press, Cambridge, US

Rochat P (2003) Five levels of self-awareness as they unfold early in life. Conscious Cogn 12:717–731

Rochat P (2005) Humans evolved to become *Homo Negotiatus* . . . the rest followed. Behav Brain Sci 28:714–715

Rochat P (2006) What does it mean to be human? J Anthropol Psychol 17:100–107

Rochat P (2007) Intentional action arises from early reciprocal exchanges. Acta Psychol 124:8–25

Rochat P (2008, in press) Others in Mind: social origins of self-consciousness. Cambridge University Press, NY

Rochat P, Callaghan T (2005). What drives symbolic development? In: Namy L (ed) Symbolic use and understanding. Lawrence Erlbaum Associates Publishers, Mahwah, NJ

Rochat P, Senders SJ (1991) Active touch in infancy: action systems in development. In: Weiss MJ, Zelazo PR (eds) Infant attention: biological contraints and the influence of experience. Ablex Publishers, Norwood, NJ, pp 412–442

Rochat P, Striano T (1999a). Social cognitive development in the first year. In: P Rochat (ed) Early social cognition. Lawrence Erlbaum Associates, Mahaw, pp 3–34

Rochat P, Striano T (1999b) Emerging self-exploration by 2 month-old infants. Dev Sci 2:206–218

Spitz RA (1965) The first year of life: a psychoanalytic study of normal and deviant development of object relations. Basic Books, New York

Stern D (1985) The interpersonal world of the infant. Basic Books, New York

Striano T, Rochat P (1999) Developmental link between dyadic and triadic social competence in infancy. Br J Dev Psychol 17:551–562

Striano T, Rochat P (2000) Emergence of selective social referencing in infancy. Infancy 2:253–264

Tomasello M (1995). Joint attention as social cognition. In: Moore C, Dunham P (eds) Join attention: its origins and role in development. Erlbaum, Hillsdale, NJ, pp 103–130

Tomasello M (1998) One child early talk about possession. In: Newman J (ed) The linguistic of giving. John Benjamins, Philadelphia, pp 349–373

Tomasello M (1999) Cultural origins of human cognition. Harvard University Press, Cambridge

Tomasello M, Call J (1997) Primate cognition. Oxford University Press, New York

Tomasello M, Farrar MJ (1986) Joint attention and early language. Child Dev 57: 1454–1463

Trevarthen C (1979). Communication and cooperation in early infancy: a description of primary intersubjectivity. In: Bullowa MM (ed) Before speech: the beginning of interpersonal communication. Cambridge University Press, New York, pp 321–347

De Waal FBM (1989) Peacemaking among primates. Harvard University Press, Cambridge, MA

Wellman HM (2002). Understanding the psychological world: developing a theory of mind. In: Goswami U (ed) Blackwell handbook of child cognitive development. Blackwell Publishing, Oxford, UK, pp 167–187

Wimmer H, Perner J (1983) Beliefs about beliefs: representation and constraining function of wrong beliefs in young children's understanding of deception. Cognition 13:103–128

Winnicott DW (1982) Playing and reality. Tavistock Publications, London/NewYork

Zahn-Waxler C, Radke-Yarrow M, Wagner E, Chapman M (1992) Development of concern for others. Dev Psychol 28:126–136

8
Little Liars: Origins of Verbal Deception in Children

Victoria Talwar[1] and Kang Lee[2]

1. Introduction

Scientists and laypersons alike have been fascinated with lying in children since the dawn of developmental psychology (Darwin 1877). This fascination stems from the fact that the seemingly innocent and poorly crafted lies told by children are kaleidoscopes from which one can glean a multitude of information about children's development including their level of moral development (Piaget 1932), their ability to understand of others' minds (Peskin 1992), personal character and integrity (Hartshorne and May 1928), delinquent tendencies (Stouthamer-Loeber 1986), and credibility as witnesses in legal cases (Bala et al. 2001; Bussey et al. 1993; Goodman et al. 2006; Lyon 2000).

The scientific study of the development of lying began at the turn of the 20th century, when developmental psychology as a field of scientific research was just being established. Recently, after having neglected the topic for more than a half century, developmental researchers with diverse theoretical orientations and research purposes are showing a renewed interest in the development of lying in children. Many developmental psychologists have converged on the subject of the development of lying for both theoretical and practical reasons. Research on children and lying is a cross-cutting field of empirical inquiry that is related to diverse areas of research and practice including cognitive evolution and development, emotion and its development, social psychology, cross-cultural study, clinical and legal practice, and moral education. The aim of this chapter is to review and evaluate the current findings about the development of verbal deception in terms of the types of lies that children tell, their motivations for lying, age-related changes in lie-telling skills, and the cognitive and social factors that are associated with lying.

2. Development of Lie-telling Behaviour

Lying is a form of verbal communication whereby the speakers make a false statement with an intent to deceive the intended recipient (Bok 1978; Chrisholm

[1]Department of Educational and Counselling Psychology, 3740 McTavish Montréal, Québec, Canada H3A 1Y2
[2]Institute of Child Study, University of Toronto, 45 Walmer Road, Toronto, Ontano, Canada M5R 2X2

and Freehan 1977; Coleman and Kay 1981; Lee 2000). Although most of our daily communications with others are truthful, lying is more than an occasional occurrence. In fact, there is evidence to suggest that adults tell lies on a daily basis (DePaulo and Kashy 1998; DePaulo et al. 1996) with a multitude of underlying motivations and rationales. Some lies are antisocial in nature. They are typically self-serving, told to protect oneself from harm or for personal gain. Some other lies are told to benefit others, although for different reasons and not always prosocial in nature. For example, individuals may tell lies to help another individual get out a difficult situation (e.g., perjury for a friend). They may lie to spare the feelings of another. They may also tell the so-called "blue lies" (Barnes 1994) to help a collective (Fu et al. in press). While there has been extensive research on adults' lie-telling behaviour (e.g. Bond et al. 1992; DePaulo and Kashy 1998; DePaulo et al. 1996; Ekman 1985; Ekman and O'Sullivan 1991), research is considerably limited on children's lying behaviour. However, in large part, it appears children are motivated to tell lies for the same reasons adults tell lies.

2.1. *Self-serving Lies*

In theory, lies for self protection or enhancement ought to begin appearing after a sense of self has emerged which is typically around 2 years of age (Lewis et al. 1989b). One must have a sense of self in order to seek to protect it or enhance it. Indeed, anecdotal evidence and observational studies suggest children start telling lies as early as 24 months of age (Bussey 1992; Darwin 1877; Leekam 1992; Newton et al. 2000; Wilson et al. 2003).

Experimental studies examining children's antisocial lies have focused mainly on preschool children (e.g., Lewis et al. 1989b; Polak and Harris 1999; Talwar and Lee 2002a) and have typically relied on a modified temptation resistance paradigm (Sears et al. 1965). That is, children are placed in room with attractive toys and instructed not to peek at or play with the toys while the experimenter is absent. However, due to the highly tempting nature of the situation, approximately 80–90% of preschool children do not abide by the experimenter's instruction and peek at or play with the toy, thus committing a transgression. When the experimenter returns, children are asked whether they had peeked or played with the toys. As this experimental paradigm mimics the situation that young children often get themselves into, the temptation resistance situation creates a naturalistic setting in which children may be motivated to lie out of fear of reprisal from the experimenter if truth is told.

Indeed, Lewis et al. (1989a) found that among the 3-year-olds who transgressed (peeking at a toy when told not to), 38% denied peeking at the toy while 62% confessed to peeking. Talwar and Lee (2002a) extended this study by including children between 3 and 7 years of age. They found that the majority of children between 4 and 7 years of age lied about peeking at a toy whereas only approximately one-third of three-year-olds did the same, a rate similar to that found by Lewis et al. (1989a). Further, when the temptation resistance paradigm is adapted for older children between 7 and 12 years of age, the majority of them lied about their peeking (Talwar et al. 2007a). Thus, from these findings taken along with those of observational studies, it appears that some children as young as 2 or 3 years of age and most children by 4 years of age do tell lies to conceal transgressions, perhaps to avoid potential punishment. It is worth noting that this significant age pattern in which 3-year-olds are relatively less inclined to lie about their transgression than older children has been observed among American, Canadian, British, Chinese, Japanese, and West African children (Lewis 1993; Polak and Harris 1999; Talwar and Lee 2002a, 2004; Talwar et al. 2002a; Xu and Lee 2007). Hence, there is convergent evidence to suggest that this age pattern is a universal phenomenon.

This universal pattern of development in lying may also apply to lies told for personal gain. In one study, for instance, children were taught to mislead a puppet about the location of a prize in order to keep the prize for themselves (Sodian et al. 1991). Children were not reliably capable of committing this deception until four years of age. In a similar study in which children could keep a sticker if they chose to misinform a puppet who was also interested in the sticker, Peskin (1992) found that 87% of five-year-olds lied about the location of a prize to the puppet, whereas only 29% of three-year-olds did so.

2.2. Lies to Conceal Another's Wrongdoing

Children may also be motivated to tell lies for others to conceal their transgressions. Wilson and Pipe (1995) had an unfamiliar adult "magician" ask 5- and 6-year-olds to conceal the adult's accidental spillage of ink on a pair of gloves. They found that the children did not spontaneously mention the accident when later interviewed. In a similar study, Pipe and Wilson (1994) had 6 and 10-year-olds participate in a "magician show" in pairs. One child was made an assistant to the magician and the other one was a spectator. During the magic show the magician spilt ink on the assistant's gloves and asked the child to keep a secret. They found 6-year-olds were significantly more likely than 10-year-olds to keep a secret about damage caused by the unfamiliar adult when asked about the event (Pipe and Wilson 1994). Of the younger children 60% revealed the truth about the accident when first interviewed while 84% of older children did so. It appears that children are generally unwilling to lie for individuals to conceal their transgressions, which is consistent with an early finding that children would not lie for another child even though the child had helped them previously (Greenglass 1972).

One possible explanation of why children refused to lie for the individuals in the above mentioned studies is that the children had hardly known those individuals and thus did not have a strong motivation to help them by lying for them. If this possibility is true, one should expect children to be inclined to lie for someone with whom they are highly familiar. The results of Bottoms et al. (2002) partially supported this explanation. In their study, while the experimenter was out of the room, they played with some toys. Their mother broke one toy and then told the child not to tell the experimenter about the breakage because she might get in trouble. It was found that older children were more likely than younger children to withhold information about their parent breaking the toys. Thus, it appears that at least older children would be willing to conceal their parents' transgression, perhaps out of loyalty to them.

However, this loyalty, if any, is rather fragile as evidenced by a recent study by Talwar et al. (2004). In their study, parents committed a minor transgression (i.e. breaking a puppet) while the experimenter was out of the room. They coached their child not to tell the experimenter that the parent had broken the puppet because they might get in trouble if the experimenter found out. Later, children were interviewed about what happened to the puppet either with or without parents being present. Regardless of the conditions, the majority of children did not conceal their parent's transgression, and told the truth about their parent's transgression. Many of the children told on their parents immediately even though the experimenter had not noticed the breakage of the puppet. Interestingly, children became more inclined to lie for their parents after it was clear to children that they themselves would not be blamed for breaking the puppet. These results taken together suggest that children will tell lies for others but may be motivated only when their decision to lie does not impinge on self-interest. In other words, self-interest rather than familiarity with an individual may be a stronger motivator for children to decide to lie for another individual.

2.3. Prosocial Lies

Prosocial lies are an interesting development in children's repertoire of social behaviours. While non-human primates have been observed to use tactical deception for self-interests, prosocial deception appears to be unique to humans. During our interaction with others, from time to time, we are required by social conventions not to speak the truth to a recipient with an intention to help the recipient (Sweetser 1987). For example, when we receive an undesirable gift from a gift-giver, social etiquette requires us not to tell the gift-giver bluntly that we do not like the gift. Instead, when asked, we are required to tell a white lie.

Evidence suggests that children may tell prosocial lies as early as they lie to conceal their own transgression. In one study, we examined children's "white lies" in children ages 3 to 7 years (Talwar and Lee 2002b). A Reverse Rouge Task paradigm was used in which the experimenter had a conspicuous mark of lipstick on the nose. The child was asked to take a picture of the experimenter, but before the picture was taken the experimenter asked, "Do I look okay?" Results showed that 89% of children between 3 and 7 years of age responded "yes" to the experimenter but later told another adult that they thought the experimenter did not look okay.

However, these children might have lied for entirely prosocial reasons. In this situation, if children told the truth that the adult did not look okay, they might face negative reactions from the adult; lying not only avoided unpleasant repercussions but also might please the adult. Thus, children had little to lose but much to gain by telling a lie in this situation. To address this issue, one needs to place children in situations in which they may choose to lie in order to protect another's feelings at the expense of their self-interests. We examined children's behaviour in such a situation using an undesirable gift-paradigm (Talwar et al. 2007b). Children played a game where they were promised a gift from a gift basket that contained a range of different toys and gifts. After the task children received an undesirable gift of soap instead of a toy and were questioned by the gift-giver about whether they liked the gift. In this situation children had to reconcile their desire for a better gift with the competing social requirement to be polite. When asked if they liked the gift, the majority of children told the gift-giver untruthfully that they like the disappointing gift despite having told their parents that they did not like the gift. However, school-aged children were more likely to tell a prosocial lie than preschool children. In a similar study conducted in China, we found while 15% of seven-year-olds told a prosocial lie, 48% of ll-year-olds did so despite the negative consequences to themselves (Popliger et al. 2007). The results from the Reverse Rouge and Undesirable Gift Paradigms taken together suggest that children may tell a white lie as young as 3 years of age but self-interests may be an important motivating factor for younger children. With increased age, however, children are able to allow prosocial motivations to override their self-interest and tell genuine prosocial lies.

2.4. Summary

Children's lie-telling appears to emerge early in the preschool years. Children's first lies appear to be motivated by self-interest and the desire to avoid punishment. As children become older they are more likely to be concerned about the feelings and needs of others and are more likely to tell lies for others. However, they are still motivated by self-interest and when these needs conflict with another's they are more likely to tell the truth or lie according to their own interests. It is not until the school-age years, that children start to tell lies for others when there may be a cost to themselves. However, in many of these cases like the situation where one receives an undesirable gift the cost is not extremely high. So like adults, children are more likely to tell lies for self-interests but will conform to social norms and etiquette to tell prosocial lies to maintain social relationships when costs to self-interest are low.

3. Children's Success at Lying

Lie-telling, whether told for personal gain or for another, may have grave consequences if the lie is detected. In order to be successful at lying, a lie-teller must maintain their expressive behaviour in a manner that is congruent with their lie. A person feigning sickness must alter their non-verbal appearance to appear ill. At the same time, the person must also avoid making other statements that are inconsistent with the lie (e.g., talking about a shopping expedition or rigorous tennis practice during the period of the reported illness). Similarly, a person verbally expressing delight at receiving an undesirable gift must continue to make positive statements about the gift in the presence of the lie recipient as well as faking positive non-verbal expressions.

Successful lying thus requires one to inhibit what are genuine verbal and nonverbal reactions to a situation. At the same time, the liar needs to use appropriate display rules for the situation and simulate verbal and nonverbal expressions as required by these rules. Both verbal and nonverbal display rules can be defined as rules governing communications between individuals for relaying information, expressing emotion, and conveying attitude (Saarni 1979). Such rules can guide individuals to modify their public expressions of private information, feeling, and attitude, and help them determine what is appropriate or inappropriate to display either verbally or nonverbally in a given social setting. Individuals are required to comply with the display rules in situations where empathy, courtesy, and customary etiquette are normally expected. In politeness situations (e.g., when receiving an undesirable gift), one is thus required not only to suppress the genuine verbal and nonverbal reaction of disappointment, but also use the appropriate display rules for the situation (e.g., expressing gratitude toward the gift giver in conjunction with the display of positive emotional expressions). In order to be successful at lying, children must co-ordinate both non-verbal and verbal expressive behaviours.

3.1. Nonverbal Expressive Behaviour and Deception Detection

Research with adults has shown that there is no typical deceptive behaviour, but some behaviours are more likely to occur than others during deception, depending on emotions experienced by the liar, the complexity of the lie, and the amount of effort the liar exerts in controlling his/her behaviour (Vrij 2002). Commonly, telling a lie evokes three common emotions: fear, guilt, and excitement or "duping delight" (Ekman 1985). Liars may be afraid of getting caught, feel guilty when they lie, or feel excited to have the opportunity to fool somebody. Theses emotions can be reflected in a lie-teller's nonverbal behaviour (e.g. gaze aversion, smiling, fidgeting) and consequently betray the lie.

It has been suggested that children may be unable to regulate their nonverbal expressive behaviour when deceiving due to their less sophisticated level of cognitive development and their inability to understand the implications of their non-verbal behaviour. Research on children's emotional regulation has found evidence to suggest that children's ability to monitor their expressive cues and their understanding that internal emotional experience and external affect expressive need not correspond to each other develops only in elementary school years (Saarni 1979, 1984). This would suggest that children's ability to control their non-verbal expressive behaviour when deceiving may not develop after preschool years.

Indeed, several studies have found that young children were poor at "feigning" dislike or liking of different stimuli (e.g.,a drink) (Feldman and White 1980;

Feldman et al. 1979; Morency and Krauss 1982). For example, Feldman et al. (1979) investigated whether school aged children could control their facial expressions while lying to others and the effectiveness of their deception. In their study, they asked 1st grader, 7th grader and college students to taste two new beverages with one tasting pleasant and the other unpleasant. For both drinks they had to convince an experimenter that both beverages tasted good or bad. Adult observers were able to discern differences in younger children's truthfulness but had difficulty doing so with 7th graders and college students. First graders did not mask their facial expressions whereas 7th graders simulated facial expressions that were consistent with their lies (e.g., smiling when drinking an unpleasant beverage). One of the major problems of this series of studies is that participants were instructed by the experimenter to "pretend" that they liked or disliked a stimulus. Children, particularly younger ones, might not have taken seriously the task given (e.g., treating the task as a game) and thus did not feel a strong need to regulate their emotional expressions. Their success at regulating emotional behaviours might be different if they had to fake liking or disliking the stimuli due to their own volition.

Indeed, research on children's emotional regulation has shown that children as young as 3 and 4 years of age are able to exercise spontaneous expressive control in a situation when they received a disappointing gift (Cole 1986). Furthermore, a handful of studies that have examined children's spontaneous lie-telling abilities have also found evidence that young children are able to avoid detection through non-verbal expressive control. For instance, Lewis et al. (1989a) examined a number of expressive behaviours displayed by 3-year-olds who chose to lie out of their own volition about their transgression (i.e., peeking at a toy) found only one small difference: lie-tellers displayed more positive behaviours than nonliars. They also found that adult raters were unable to distinguish between the lie- and truth-tellers based on the children's nonverbal behaviours. Furthermore, Talwar and Lee (2002a) conducted in-depth analyses of the children's non-verbal behaviours between 3 and 7 years of age and also found that lie-tellers were almost indistinguishable from truth-tellers in terms of their non-verbal behaviours. Only a small difference was found: more lie-tellers showed big smiles than the truth-tellers, while more truth-tellers were likely to have a relaxed mouth. When video-clips were shown to adults (university students, parents, police officers, and custom officers), they performed at or near chance levels in their detection of children's lies (Leach et al. 2004; Talwar and Lee 2002a). Thus, it appears children as young as 3 years of age are successful at nonverbal leakage control when lying to conceal their own transgressions. In situations where children are lying to conceal their own transgression, they have to suppress negative emotions such as fear of being caught in order to avoid detection.

However, in situations of prosocial lying for politeness purposes, children are required not only to suppress negative emotions such as disappointment but also substitute it by showing positive expressive displays appropriate for the situation. Children may find it harder to perform such a dual task. Indeed, Talwar and Lee

(2002b) found that when white lie-tellers who lied about the appearance of an adult displayed fewer big smiles and appeared more serious and concerned than control children. The latter appeared confident and relaxed. However, overall, white-lie tellers and truth-tellers were highly similar both in terms of the total number of positive and negative expressive behaviours and in terms of adults' abilities to differentiate between children when viewing video clips of their expressive behaviours. Similarly, Talwar et al. (2007b) found children between 3 and 11 years of age who told a white lie about liking a disappointing gift were able to suppress negative expressions and simulate appropriate positive behaviours. In particular, children masked their genuine negative emotions when the gift-giver was present but did not when she was absent. When alone with the disappointing prize they showed disappointment but this negativity was quickly replaced with smiling in front the experimenter.

Thus, it appears that children are successful in concealing their deception in terms of their non-verbal behaviour in a variety of deceptive situations. While some deceptive signs were revealed when children's expressive behaviour was coded by expert coders, these differences clearly did not help naïve adults distinguish lie-tellers from truth-tellers.

3.2. Verbal Expressive Behaviour and Deception Detection

However, non-verbal behaviour is not the only way children's lies can be detected. Children may be difficult to detect in terms of their non-verbal behaviour but what about their verbal behaviour? Equally important in ensuring the success of a lie, a liar must be able to maintain their lie verbally, avoiding inconsistencies between their initial and subsequent false statements. The ability to maintain one's lie and avoid inconsistency is referred to in the literature as semantic leakage control.

Observational data in daily life settings has revealed that four-year-olds' lies typically take the form of one word responses rather than the more sophisticated elaborations of older children and adults (Bussey 1992). These observations suggest that younger children may be less sophisticated at semantic leakage control. Talwar and Lee (2002a) tested this possibility by asking children follow-up questions after they had denied peeking at a toy. One of the follow-up questions was concerned with the identity of the toy. Because there were obvious clues about the toy's identity without looking at it, children must not readily report the identity of the toy if they had claimed they had not peeked at the toy. A significant age difference was found. Whereas most preschoolers blurted out the identity of the toy and thus implicated them in their transgression and lying, only half of six and seven-year-olds did so. The other half feigned ignorance about the identity of the toy (Talwar and Lee 2002a). In another study, we saw this age trend continue with children between 7 and 11 years of age with most of the 11-year-olds strategically faking ignorance about answers to follow-up questions (Talwar et al. 2007a). Thus, it appears that as young children are limited in their semantic leakage control and are not skilled lie-tellers, as they get older children become more sophisticated at concealing their lies verbally.

3.3. *Summary*

Thus far research on children's lie-telling behaviour suggests children as young as 3 years of age are skilled at exercising nonverbal leakage control, but they have poor semantic leakage control under 6 years of age. However, in everyday situations we rarely attend to either only non-verbal or only verbal behaviour but rather to the person's entire expressive behaviour. Thus, if we take the findings from both children's ability to regulate their non-verbal and verbal behaviour, it appears that children under 6 years of age overall have yet to develop all the skills necessary to deceive adults successfully. However, these abilities may develop rapidly as children reach school age (Talwar and Lee 2002a; Talwar et al. 2007a).

4. Cognitive and Social Factors Related to Lie-telling

The development of children's abilities to deceive successfully may be due to both their social and cognitive development. A few studies have begun to examine cognitive and social factors contributing to the development of lying.

4.1. *Theory of Mind Understanding*

Theory of mind refers to a collection of understandings about our own and others' mental life such as emotion, desire, knowledge, and belief. One of the key elements of theory of mind related to lying is false belief understanding, or the understanding that another individual may have a false belief about the true state of affairs that differs from our own. It has been suggested that children's false belief understanding might be related to their lie-telling ability because lying, by definition, is to instil false belief into the mind of a recipient (Chandler et al. 1989; Polak and Harris 1999; Talwar and Lee 2002a). A precondition to tell a lie and tell it successfully is that a lie-teller must understand that they have privileged knowledge about the state of affairs (e.g., whether they have peeked at a toy) to which the lie-recipient does not have access. For this reason, the ability to deceive has been seen as one of indicators regarding children's level of theory of mind understanding in general and false belief understanding in particular (Sodian 1991). In a sense, lying is theory of mind in action.

Extensive research has shown that whereas most of children above 4 years of age are able to understand that others may have a false belief about the true state of affairs, many 3-year-olds do not (for a review, see Wellman et al. 2001). These findings suggest that young children may be incapable of intentionally deceiving. Yet, studies suggest children as young as 3 are capable of misleading another person. In a controversial study, Chandler et al. (1989) examined whether young children are capable of initiating a deceptive action. They had children between 2 and 4 years of age hide a treasure in 1 of 4 containers so that the adult would not be able to locate it. They found that children used a variety of deceptive strategies including withholding evidence and lying. From these findings, Chandler et al. (1989) claimed that 3-year-olds children engage in deception with the intention to create a "false belief" in others. This finding remains controversial (Sodian 1991; Sodian et al. 1991) as subsequent studies showed that 3-year-olds as a group did not appear to show the ability to use deceptive acts spontaneously.

One possible reason for this discrepancy may be that 3-year-olds are not homogenous. It is well established that there exists considerable variability in 3-year-olds' false belief understanding. Whether 3-year-olds act deceptively may be related to whether they have acquired such understanding. To test this hypothesis, Polak and Harris (1999) and Talwar and Lee (in press) used the temptation resistance paradigm to elicit spontaneous lying behaviour in children between 3 and 7 years of age. They also assessed and the children's theory of mind understanding using standard false belief measures. In the context of the temptation resistance paradigm, it has been suggested that children's false denials that they peeked at or played with a forbidden toy only require the child to represent a belief that is different from the true state of affairs (Polak and Harris 1999; Talwar and Lee 2002a; Talwar et al. 2007a). Therefore, to falsely deny requires understanding requires the representation of a first-order false belief (referred to as the ToM1 Hypothesis) and thus whether children would lie to deny their peeking or playing should be related to their first order false understanding scores. The results by Polak and Harris (1999) and Talwar and Lee (in press) confirmed the ToM1 Hypothesis. They found that children's first-order false belief understanding was significantly correlated to their initial false denials even after the effect of age had been partialed out.

However, they found that first-order false belief understanding was not significantly related to children's ability to feign ignorance. They suggested that feigning ignorance to follow-up questions (e.g., "What do you think the toy is?") requires children to be able to represent second-order mental states. For example, after peeking, children can take the experimenter's perspective and first assume that the experimenter thinks they have no knowledge of the answer because they have said they have not peeked at the toy (a false belief). Given this false belief, children need to reason what the experimenter expects them not to know about the identity of the toy (a second belief). Thus, after having lied about peeking at the forbidden toy, when asked the identity of the toy, children must feign ignorance to avoid violating the experimenter's expectation. In other words, children must be able to represent a belief about the experimenter's false belief to be able to maintain consistency between the initial lie and the subsequent statement. Thus, the inability of preschool children to feign ignorance in subsequent verbal statements found in Polak and Harris (1999) and Talwar and Lee (2002a) may be due to the fact that the children in these two studies have not acquired the ability to represent second-order beliefs (the ToM2 Hypothesis).

This suggestion is consistent with previous research that has shown that second-order mental state understanding begins to emerge only around 6 years of age and undergoes steady development well into adolescence (Hogrefe et al. 1986). Further, Banerjee and Yuill (1999) found that children who passed second-order belief tests were more likely to suggest story protagonists make false claims so as to present themselves in a positive light to others. To test the ToM2 hypothesis, Talwar et al. (2007a) assessed the lying behaviour and second-order false belief understanding of children between 7 and 12 years of age. They found that older children were indeed more likely to feign ignorance in follow-up questions than younger children and their success was significantly correlated with performance on second-order belief tasks after the effect of age was partialed out.

4.2. Executive Functioning

Some evidence exists that children's lie-telling behaviour also might be related to executive functioning, highlighting the potential role of prefrontal maturation. Executive function has been defined as higher-order psychological processes involved in goal-oriented behaviour under conscious control (Zelazo and Muller 2002). Executive functioning encompasses a collection of cognitive skills including self-regulation, inhibitory control, planning, attentional flexibility, strategy employment and influence (Welsh et al. 1991; Zelazo et al. 1997). Executive functioning skills have been shown to emerge in late infancy and develop during the childhood years (Welsh and Pennington 1988; Zelazo and Muller 2002), a time when researchers have noted increases in lie-telling skill (e.g., Polak and Harris 1999; Talwar and Lee 2002a). In particular, it has been suggested that inhibitory control and working memory may be directly related to children's deception (Carlson and Moses 2001; Carlson et al. 1998), and such a relation also might be mediated by theory of mind development (Carlson et al. 2002; Sabbagh et al. 2006). Inhibitory control is the ability to suppress interfering thought processes or actions (Carlson et al. 2002) and working memory is a system for temporarily holding and processing information in the mind (Baddeley 1986). When lying, children must suppress their reporting of the transgression that they wish to conceal and represent and utter the false information that differs from reality (Carlson et al. 1998, 2002). Additionally, in order to maintain their lies, children must inhibit those thoughts and statements that are contrary to their lie and would reveal their transgression while maintaining in their memory the contents of their lie. Thus, children must hold conflicting alternatives in their mind (i.e., what they really did/thought and what they said they did/thought).

There are only two studies that have examined the relation between children's executive functioning and deceptive behaviour (Carlson et al. 1998; Talwar and Lee in press). Carlson et al. (1998) found that preschool children who experienced difficulty with executive functioning tasks, especially those that require a high level of inhibitory control, demonstrated difficulties with physical deception (i.e., pointing). Although Carlson et al. (1998) did not explicitly examine lie-telling behaviour, their results seem to suggest that children may also have difficulties with lying if they lack advanced executive functioning skills, particularly in terms of inhibitory control and working memory. To address these possibilities, Talwar and Lee (in press) used the temptation resistance paradigm with children between 3 and 8 years of age. They found that a significant relation between children's executive functioning and their lying behaviour. Children with higher stroop task scores were more likely to lie. This significant relation is expected because when children were asked if they peeked, they had to suppress the reporting of the transgression that they wished to conceal, and represent and utter the false information that differs from reality. The inhibitory control that is needed to tell such a lie may be the same executive functioning skills that are involved in performing the stroop task. The stroop task also involves working memory (see Carlson and Moses 2001). It is thus possible that working memory may also play a role in children's decision to lie. Telling a lie may require the dual ability to remember the rule being violated and inhibit reporting of the transgression that they wish to conceal. This finding suggests that the development of children's lie-telling abilities is not only related to children's theory of mind abilities but also to their executive functioning skills. This finding may not be surprising because research has consistently shown a significant relationship between children's theory of mind understanding and their executive functioning (e.g., Carlson et al. 2002; Hughes 1998; Perner et al. 2002). Interestingly, dual executive demand tasks like the stroop task used in this study have been found to be more strongly predictive of theory-of-mind capabilities than working memory or inhibitory tasks alone (e.g. Carlson and Moses 2001; Hala et al. 2003). Given this finding, it may be that the combination of inhibitory control and working memory may be crucial not only for ToM reasoning but also for lying which requires children to put their theory-of-mind knowledge into action.

4.3. Conceptual and Moral Understanding of Lying and Honesty

As lying is generally considered a socially and morally negative behaviour, an important question is whether children's lying behaviour is linked to their knowledge about the concept of lying and its moral implications. One may be tempted to argue that once one knows that lying is bad, then our behaviour falls in line and we remain honest. Yet, a moment's reflection will reveal the glaring simplicity of such an argument. In fact, adults tell lies despite knowing that lie-telling is considered morally wrong under most circumstances and act to appear moral (Batson and Thompson 2001).

Developmental research has demonstrated that a child's conceptual and moral understanding of lie- and truth-telling emerges early in the preschool years and develops rapidly throughout the school years (Bussey 1992, 1999; Peterson et al. 1983; Piaget 1932; Siegal and Peterson 1998; Talwar et al. 2002a); see (Lee 2000 for a review). As young as 3 years of age, children already have a rudimentary concept of lies that are told for antisocial purposes and they evaluate such lies negatively. With increased age, children begin to differentiate antisocial lies from honest mistakes, guesses, exaggerations, and eventually sarcasm and irony. Children also gradually take into consideration the social context in which lies are told and the intention of the lie-teller when evaluating lies. Overall, by early adolescence, children's conceptual and moral understanding of lying and truth-telling becomes comparable to adults.

However, there has been limited research to examine the relationship between children's conceptual and moral understanding and their actual lying behaviour. Talwar et al. (2002a) found no relationship between children's actual lie- and truth-telling behaviour and their conceptual and moral understanding of lies. The majority of children who reported that lying to conceal a transgression was bad, could correctly identify such a lie, and recommended others to tell the truth. Nevertheless, most of them told lies to conceal their own transgressions. Although children's moral evaluations did not predict their actual behaviour, interestingly children's promising to tell the truth did influence their behaviour. Children who promised to tell the truth were less likely to lie about their peeking behaviour. Another study by Talwar et al. (2004) found a significant but modest correlation between children's conceptual and moral understanding and their lie-telling behaviour to conceal a parent's transgression. Also, promising to tell the truth decreased the likelihood of children's lie-telling for their parents (Talwar et al. 2004).

In both studies children's conceptual and moral understanding was assessed via tasks commonly used by legal professionals in the court (Bala et al. 2001; Lyon 2000; Talwar et al. 2004). These tasks are typically very brief and only assessed whether children had a minimal understanding of lying and truth-telling. As a result, there was low variability in children's scores which may have obscured a genuine relationship between children's developing conceptual and moral knowledge and their lie-telling behaviour. To address these problems, Talwar and Lee (in press) used a battery of measures that assessed the conceptual and moral understanding of lying and honesty in children between 3 and 8 years of age, along with the temptation resistance paradigm for assessing lying behaviour. Consistent with previous studies, children's concept classification scores in this study increased with age (Bussey 1992, 1999; Lee 2000; Peterson et al. 1983; Siegal and Peterson 1998; Talwar et al. 2002a). Although a much more comprehensive measure of children's conceptual knowledge was used in the present study, the findings were similar to the two other studies that have examined the relationship between children's conceptual understanding and their actual lying behaviour (Talwar and Lee 2002a, 2004). Perhaps this is not surprising given that adults know what a lie is but still lie on a day-to-day basis (DePaulo and Kashy 1998).

Similar to previous studies (e.g., Bussey 1992, 1999; Piaget 1932; Siegal and Peterson 1998), children's age was significantly related to their evaluations of the moral stories. When making moral judgments, younger children were more likely to attend to factors of factuality and promising than older children. However, they were less likely to take motivation into consideration when compared to older children. Thus, younger children paid more attention to the factuality of a statement and the adherence or violation of rules (e.g., promise-keeping or breaking), to make their moral evaluations, whereas older children considered the character's intention to deceive itself to make their evaluations. These findings are in keeping with previous studies which have found that while children as young as 4 years of age can make basic distinctions between lies and truth, their moral understanding of lies develops overtime with younger children being more influenced by factuality of statements and external factors while older children are more influenced by intentions and internal factors (e.g., Bussey 1992, 1999; Peterson et al. 1983; Piaget 1932).

More importantly, Talwar and Lee (under review) found that children's lying behaviour was related to their moral evaluations of truthful and untruthful statements. More specifically, children who admitted their transgression were more likely to value truthfulness and give it higher ratings regardless of the situation. In contrast, children who chose to lie tended not to have stringent views about the need to be truthful. These results suggest that children who hold more relativist views about the moral implication of lying might be more inclined to tell lies, whereas those who held more stringent moral views about lying are more likely confess. Another noteworthy finding is that in comparison with the liars and confessors, those children who abided by the experimenter's instruction and did not peek (non-peekers) gave the most positive ratings for stories where the protagonist kept a promise and the most negative ratings when the story protagonist failed to keep a promise. One possible explanation is that non-peekers were the most concerned about rules and adherence to them. They might have taken the experimenter's instruction about not peeking at the forbidden toy more seriously than the lie-tellers and confessors, and this concern was strong enough to motivate them to resist the high level temptation to peek in the current procedure. Thus, while children's ability to classify whether a statement was a lie was not related to their behaviour, their perceptions of the acceptability of such statements were significantly related to their behaviour.

It should be noted that the above studies have focused on lying to conceal transgressions. When children tell lies for prosocial purposes, the relations between their behaviour and moral knowledge tend to be stronger. For example, in one study in which Chinese children between 7 and 9 years of age were given an undesirable gift, children's moral evaluations of prosocial lies were found to be significantly related to their actual prosocial lie-telling behaviour with an effect size of more than 30% (Xu et al. under review). Specifically, children who evaluated prosocial lies less negatively were more likely to tell a prosocial lie to protect the feelings on another.

Fu et al. (in press) also studied the relation between Chinese children's conceptual knowledge of lying and children's actual lying for the benefit of a collective. Seven, 9-, and 11-year-old Chinese children were placed in a real life situation where they had to decide whether to lie to help a collective (behavioral measure). Children's willingness to endorse lying or truth-telling that benefits a collective but at the same time harms an individual was evaluated with stories depicting hypothetical situations where story characters also faced the dilemmas of telling a lie for the collective versus the truth for an individual and vice versa. Results demonstrated that as age increased, children became more inclined to endorse lying for the collective, reflecting the increase influence of collectivism promoted in the Chinese society on their moral valuations. Also, the children became increasingly willing to tell lies for their group themselves. Further, children's endorsement about blue lies in hypothetical situations predicted their actual behaviour with an effect size of more than 20%.

The stronger moral knowledge-behavior relation found in the above two studies than found in previous studies may be explained by the types of lies under question. In generally, social and moral norms tend to guide our moral value judgments more strongly than social-situational factors, whereas social-situational factors tend to guide our actions more strongly than social and moral norms. When faced with making a decision of whether to tell a lie about a transgression, children's moral values may compel them to tell the truth, while the social-situational factors of self-protection entice them to tell a lie. This creates a conflict between moral norms and situational factors. Conversely, when children are faced with whether to tell a white lie, or lie to help a collective, no such conflict exists: both social and moral values and social-situational factors all support them in telling such a lie. As a consequence, the relation between children's evaluation of lies and their behaviours is stronger when lying appears to serve prosocial purposes than when lying is self-serving.

4.4. Summary

Evidence to date suggests that children's lying behaviour is significantly related to their theory of mind understanding, executive functioning, and moral judgments of lying, but not to their conceptual knowledge about lying. These significant relations, however, vary in terms of strength depending on the type of lies that children tell and how elaborated the lies are, reflecting the differential roles that various social and cognitive factors play in the development of lying.

5. Conclusions and Future Directions

Research on the development of lying dates back to the turn of the 20th century (Darwin 1877; Hall 1891; Stern and Stern 1909). Despite its early beginnings, most of the empirical work on the topic has been conducted in the last three decades. To date, we have learned a great deal about the development of lying in children. It is now well established that lying starts early at 2–3 years of age and develops rapidly. Children's ability to lie improves with age and by middle childhood children have become skilled lie-tellers, able to control both their non-verbal and verbal behaviour to avoid detection. Children tell lies not only to benefit themselves but also to benefit others. However, children's lie-telling or truth-telling behaviour is often motivated by self-interest. Yet, there are times when children will lie to benefit another despite cost to themselves. Although lying has been linked to behavioural misconduct and juvenile delinquency in late childhood

(Gervais et al. 2000; Stouthamer-Loeber and Loeber 1986), early lying behaviour reflects sophistication in young children's social and cognitive development.

The existing findings begin to provide a developmental picture of the origins of lying in children. Children's lying appears to progress through three levels (see also Polak and Harris 1999; Talwar et al. 2007a). First, children's "primary lies" begin around 2 to 3 years of age when children are first able to deliberately make factually untrue statements. Children's first falsehoods are often linked to situations of rule violations and children's attempt to avoid incrimination, protect self interests, or present themselves in a more positive light (Newton et al. 2000; Wilson et al. 2003). Given the fact that genuine lies told by children in later childhood tend to serve similar functions, such early falsehood may be a rudimentary form of intentional verbal deception. However, at this age children's lies are still infrequent (Newton et al. 2000; Wilson et al. 2003) with approximately half of 3 year-olds lying about their transgressions, while the remainder tend to be honest and confess their transgression when asked by adults (e.g., Lewis et al. 1989a; Polak and Harris 1999; Talwar and Lee 2002a).

The second level, "secondary lies", reflects a significant shift that takes place between 3 and 4 years of age (Chandler et al. 1989; Peskin 1992; Polak and Harris 1999). At and after 4 years of age, the majority of children will readily tell a lie to conceal their own transgression. Children's acquisition of first-order belief understanding and executive functioning abilities (particularly in the realms of inhibitory control and working memory) may play an important role in children's progression from the first to the second level. Perhaps relying on these two important cognitive abilities, children begin to regulate successfully their verbal and nonverbal behaviours to appear honest (Talwar and Lee 2002a). However, many children at the secondary level appear to have difficulty at semantic leakage control and thus fail to sustain their initial lie Their subsequent statements following an initial false statement tend to be inconsistent with the initial lie and thus make their deception readily detected by naïve adults.

The third level, "tertiary lies", emerges around 7 to 8 years of age. At this level, children become gradually more and more sophisticated at semantic leakage control. Children will tell a deliberate lie while ensuring that their subsequent statements do not contradict the initial lie and thus make their statements difficult to distinguish from statements made by a non-liar. Children's second-order belief understanding appears to play an important role in the transition from the secondary to the tertiary level. This is perhaps because such understanding "allows intentional social coordination to occur" (Perner 1988, p. 272) such that children can reason about complex interactions between mental states involved in sustaining a lie and act appropriately.

Despite the advancement in the last 100 years in terms of our knowledge about the origin of lying in children, many questions remain unanswered. For example, although observational data suggests children as young as 2 years of age make deliberate false statements, it is unclear whether such statements are a form of word play, wish fulfillment, or genuine deception (i.e., statements made with an intent to instil false belief into the mind of the recipient). Another question revolves aroud whether it is reasonable to assume that lying is especially suited to serve adaptive functions for children because of their lack of physical strength and social power (Bok 1978). Existing evidence indeed is consistent with this assumption: children's lies are generally motivated by self-interest. However, being adaptive by definition requires one to be flexible in one's actions when facing different situations. In some situations, lying may be called for but in other honesty may be a better strategy. No evidence exists as to whether children are capable of making strategic decisions about when to lie and when to tell the truth, and whether they are able to make cost-benefit analysis before making such decision.

A related matter is that children in different cultural environments may have to be adaptive to their unique environmental demands, which may sometimes call for lying in one cultural environment but not in another. Although there has been extensive research on cross-cultural differences in children's moral judgments of lying and honesty (e.g., Fu et al. 2007; Lee et al. 1997; see Lee 2000, and Lee and Evans in press, for reviews), limited research has been conducted to examine cultural differences in lying behaviour. Existing studies on lying to conceal one's transgression in different cultures have shown a universal pattern of development. This is not surprising because research on children's moral judgments of lying to conceal transgression has also revealed similar developmental patterns. Cross-cultural differences in lying behaviour, if any, will most likely emerge when different cultures attach different moral values to lying or honesty (e.g., lying to help one's collective: Fu et al. 2007, in press). To address this issue, children from different cultures must be examined with the use of the same paradigms.

Last but not least, to date, only a few social and cognitive factors have been examined with regard to their linkages to the development of lying. A host of variables that are potentially related to children's lying behaviour have yet to be explored. These variables include children's general intellectual abilities which some (e.g., Lewis 1993) have suggested may be related to young children's ability to lie. Another factor is parenting styles and practices as some have suggested that a harsh disciplinary style may be related directly to the development of lying to conceal one's own transgression (Lewis 1993; Stouthamer-Loeber 1986) because it is adaptive to resort to lying to avoid severe punishment. Also, parenting factors have been also been suggested to be related indirectly to greater levels of parenting stress and single parent family structure through less adaptive parenting (Cole and Mitchell 1998). The relation between children's lying and their emotional and personality factors (e.g., empathy, Machiavellian tendencies) also have never been examined. Future studies that explore the role of these factors along with those studying children's strategic lying behaviour and related cultural differences should provide a comprehensive picture about how lying emerges and development in children, a question that has fascinated laypeople and scientists alike for centuries.

References

Baddeley AD (1986) Working memory. Claredon Press, Oxford, MA

Bala N, Lee K, Lindsay RCL, Talwar V (2001) A legal & psychological critique of the present approach to the assessment of the competence of child witnesses. Osgoode Hall Law J 38:409–451

Banerjee R, Yuill N (1999) Children's understanding of self-presentational display rules: Associations with mental understanding. British Journal of Developmental Psychology 17:111–124

Barnes JA (1994) A pack of lies: towards a sociology of lying. Cambridge University Press, Cambridge, UK

Batson CD, Thompson ER (2001) Why don't moral people act morally? Motivational considerations.current directions in psychological. Science 10:54–57

Bok S (1978) Lying: moral choice in public and private life.Quartet Books, London

Bond CF Jr, Omar A, Pitre U, Lashley BR, Skaggs LM, Kirk CT (1992) Fishy-looking liars: deception judgement from expectancy violation. J Pers Soc Psychol 63:969–977

Bottoms BL, Goodman GS, Schwartz-Keeney BA, Thomas SN (2002) Understanding children's use of secrecy in the context of eyewitness reports. Law Hum Behav 26: 285–314

Broomfield KA, Robinson EJ, Robinson WP (2002) Children's understanding about white lies. Br J Dev Psychol 20:47–65

Bussey K (1992) Lying and truthfulness: children's definitions, standards, and evaluative reactions. Child Dev 63:129–137

Bussey K (1999) Children's categorization and evaluation of different types of lies and truths. Child Dev 70:1338–1347

Bussey K, Lee K, Grimbeek EJ (1993) Lies and secrets: implications for children's reporting of sexual abuse. In: Goodman GS, Bottoms BL (eds) Child victims, child witnesses: understanding and improving testimony. Guildford, New York, pp 147–168

Carlson SM, Moses LJ (2001) Individual differences in inhibitory control and children's theory of mind. Child Dev 72:1032–1053

Carlson SM, Moses LJ, Breton C (2002) How specific is the relation between executive functioning and theory of mind? Contribution of inhibitory control and working memory. Infant Child Dev 11:73–92

Carlson SM, Moses LJ, Hix HR (1998) The role of inhibitory control in young children's difficulties with deception and false belief. Child Dev 69:672–691

Chandler M, Fritz AS, Hala S (1989) Small-scale deceit: deception as a marker of two-, three-, and four-year-olds' early theories of mind. Child Dev 60:1263–1277

Chrisholm RM, Freehan TD (1977) The intent to deceive. J Philos 75:143–59

Cole K, Mitchell P (1998) Family background in relation to deceptive ability and understanding of the mind. Soc Dev 7:181–197

Cole PM (1986) Children's spontaneous control of facial expression. Child Dev 57: 1309–1321

Coleman L, Kay P (1981) Prototype semantics: the English word *lie*. Language 1:26–44

Darwin C (1877) A biographical sketch of an infant. Mind 2:285–294

DePaulo BM, Kashy DA (1998) Everyday lies in close and casual relationships. J Pers Soc Psychol 74:63–79

DePaulo BM, Kashy DA, Kirkendol SE, Wyer MM, Epstein JA (1996) Lying in everyday life. J Pers Soc Psychol 70:979–995

Ekman P (1985) Telling lies: clues to deceit in the marketplace, politics, and marriage. W. Norton, New York

Ekman P, O'Sullivan M (1991) Who can catch a liar? Am Psychol 46:913–920

Feldman R, Jenkins L, Popoola O (1979) Detection of deception in adults and children via facial expression. Child Dev 50:350–355

Feldman RS, White JB (1980) Detecting deception in children. J Commun 30:121–128

Fu G, Xu F, Cameron CA, Heyman G, Lee K (2007) Cross-cultural differences in children's choices, categorizations and evaluations of truths and lies. Dev Psychol 43:278–293

Fu G, Wang L, Evans A, Lee K (in press) Lying in the name of collective good: a developmental study. Dev Sci (in press)

Gervais J, Tremblay RE, Desmarais Gervais L, Vitaro F (2000) Children's persistent lying, gender differences, and disruptivebehaviours: a longitudinal perspective. Int J Behav Dev 24:213–221

Goodman GS, Myers JEB, Qin J, Quas JA, Castelli P, Redlich AD, Rogers L (2006) Hearsay versus children's testimony: effects of truthful and deceptive statements on juror's decisions. Law Hum Behav 30:363–401

Greenglass ER (1972) Effects of age and prior help on "altruistic lying." J Genet Psychol 121:303–313

Hala S, Hug S, Henderson A (2003) Executive functioning and theory of mind in preschool children: two tasks are harder than one. J Cogn Dev 4:275–298

Hall GS (1891) The psychology of conjuring deception. The American Journal of Psychology 11:439–510

Hartshorne H, May MS (1928) Studies in the nature of character: vol. I. Studies in deceit. Macmillan, New York

Hogrefe GJ, Wimmer H, Perner J (1986) Ignorance versus false belief: a developmental lag in attribution of epistemic states. Child Dev 57:567–582

Hughes C (1998) Finding your marbles: does preschoolers strategic behaviour predict later understanding of mind? Dev Psychol 34:1326–1339

Leach A, Talwar V, Lee K, Bala N, Lindsay RCL (2004) Intuitive lie detection of children's deception by law enforcement officials and university students. Law Hum Behav 28:661–685

Lee K (2000). The development of lying: how children do deceptive things with words. In: Astington JW (ed) Minds in the making. Blackwell, Oxford, pp 177–196

Lee K, Evans A (in press) Development of social pragmatics of deceptive communication. In: Matsui T (ed) Pragmatics and theory of mind. John Benjamins, New York

Lee K, Cameron CA, Xu F, Fu G, Board J (1997) Chinese and Canadian children's evaluations of lying and truth-telling. Child Dev 64:924–934

Leekam S (1993). Children's understanding of mind. In: Bennett M (ed) The development of social cognition: the child as psychologist. Guilford Press, New York, pp 26–61

Lewis M (1993). The development of deception. In: Lewis M, Saarni C (eds) Lying and deception in everyday life. Guilford Press, New York, pp 90–105

Lewis M, Stanger C, Sullivan MW (1989a) Deception in 3-Year-Olds. Dev Psychol 25: 439–443

Lewis M, Sullivan MW, Stanger C, Weiss M (1989b) Self development and self-conscious emotions. Child Dev 60:146–156

Lyon TD (2000) Child witnesses and the oath: empirical evidence. South Calif Law Rev 73:1017–1074

Morency NL, Krauss RM (1982). Children's nonverbal encoding and decoding affect. In: Feldman RS (ed) Nonverbal communication in children. Springer-Verlag, New York, pp 181–201

Newton P, Reddy V, Bull R (2000) Children's everyday deception and performance on false-belief tasks. Br J Dev Psychol 18:297–317

Orcutt HK, Goodman GS, Tobey AE, Batterman-Faunce JM, Thomas S (2001) Detecting deception in children's testimony; Factfinders' abilities to reach the truth in open court and closed-circuit trails. Law Hum Behav 25:339–372

Perner J (1988). Higher-order beliefs and intentions in children's understanding of social interaction. In: Astington JW, Harris PL, Olson DR (eds) Developing theories of mind. Cambridge University Press, Cambridge, NY, pp 271–294

Perner J, Lang B, Kloo D (2002) Theory of mind and self-control: more than a common problem of inhibition. Child Dev 73:752–767

Peskin J (1992) Ruse and representations: on children's ability to conceal information. Dev Psychol 28:84–89

Peterson CC, Peterson JL, Seeto D (1983) Developmental changes in ideas about lying. Child Dev 54:1529–1535

Piaget J (1932/1965) The moral judgment of the child. Free Press, New York

Pipe M, Wilson JC (1994) Cues and secrets: Influences on children's event reports. Dev Psychol 30:515–525

Polak A, Harris PL (1999) Deception by young children following noncompliance. Dev Psychol 35, 561–568

Popliger M, Talwar V, Lee K, Xu F, Fu G (2007) The development of white lie-telling among children 3- to 5-years. Presented at Society for Research in Child Development Biennial Meeting, Boston, MA

Saarni C (1979) Children's understanding of display rules for expressivebehaviour. Dev Psychol 15:424–429

Saarni C (1984) An observational study of children's attempts to monitor their expressive behaviour. Child Dev 55:1504–1513

Saarni C, von Salisch M (1993). The socialization of emotional dissemblance. In: Lewis M, Saarni C (eds) Lying and deception in everyday life. Guilford Press, New York, pp 106–125

Sabbagh MA, Xu F, Carlson SM, Moses LJ, Lee K (2006) The development of executive functioning and theory of mind. Psychological Science 17:74–81

Sears R, Rau L, Alpert R (1965) Identification and child rearing. John Wiley, New York

Siegal M, Peterson CC (1996) Breaking the mold: a fresh look at children's understanding of questions about lies and mistakes. Dev Psychol 32:322–334

Siegal M, Peterson CC (1998) Preschooler's understanding of lies and innocent and negligent mistakes. Dev Psychol 34:332–342

Sodian B (1991) The development of deception in young children. Br J Dev Psychol 9:173–188

Sodian B, Taylor C, Harris P, Perner J (1991) Early deception and the child's theory of mind: false trails and genuine markers. Child Dev 62:468–483

Stern C, Stern W (1909) The origins of lying and its prevention. Reproduced in: Stern C, Stern W, Lamiell JT (Trans) (1999) Recollection, testimony, and lying in early childhood. American Psychological Association, Washington DC

Stouthamer-Loeber M (1986) Lying as a problem behaviour in children: a review. Clin Psychol Rev 6:267–289

Stouthamer-Loeber M, Loeber R (1986) Boys who lie. J Abnorm Child Psychol 14: 551–564

Sweetser E (1987). The definition of "lie". An examination of the folk models underlying a semantic prototype. In: Hollard D, Quinn N (eds) Cultural models in language and thought. Cambridge University Press, New York, pp 43–66

Talwar V, Lee K (2002a) Development of lying to conceal a transgression: children's control of expressive behaviour during verbal deception. Int J Behav Dev 26:436–444

Talwar V, Lee K (2002b) Emergence of white lie-telling in children between 3 and 7 years of age. Merrill-Palmer Quart 48:160–181

Talwar V, Lee K (2004) Maltreated children's truth- and lie-telling behaviour. Paper presented at the American Psychological Law Society Annual Conference, Scottsdale, AZ

Talwar V, Lee K (under review) Social and Cognitive correlates of children's lying behavior. Child Dev

Talwar V, Lee K, Bala N, Lindsay RCL (2002a) Children's conceptual knowledge of lie-telling and its relation to their actual behaviours: implications for court competence examination. Law Hum Behav 26:395–415

Talwar V, Lee K, Murphy S, Griffin K (2002b) Conceptual understanding, theory of mind and executive functioning: Relation to lie-telling behaviour in children. *The International Society for behavioural Development Conference*

Talwar V, Murphy S, Samonte S (2002c) Family correlates to children's lie-telling behaviour. Poster presented at the International Society for behavioural Development Conference, Ottawa, ON

Talwar V, Lee K, Bala N, Lindsay RCL (2004) Children's lie-telling to conceal a parent's transgression: legal implications. Law Hum Behav 28:411–435

Talwar V, Gordon H, Lee K (2007a) Lying in the elementary school: verbal deception and its relation to second-order belief understanding. Dev Psychol 43:804–810

Talwar V, Murphy S, Lee K (2007b) White lie-telling in children. Int J Behav Dev 31:1–11

Vrij A (2002). Telling and detecting lies. In: Brace N, Westcott HL (eds) Applying psychology. Open University, Milton Keynes, pp 179–241

Wellman HM, Cross D, Watson J (2001) Meta-analysis of theory-of-mind development: the truth about false belief. Child Dev 72:655–584

Welsh MC, Pennington BF (1988) Assessing frontal lobe functioning in children: views from developmental psychology. Dev Neuropsychol 4:131–149

Welsh MC, Pennington BF, Groisser DB (1991) A normative-developmental study of executive function: a window on prefrontal function in children. Dev Neuropsychol 7:131–149

Wilson AE, Smith MD, Ross HD (2003) The nature and effects of young children's lies. Soc Dev 12:21–45

Wilson JC, Pipe M (1995) Children's disclosure of secrets: Implications for interviewing. In: Davies G, Lloyd-Bostock S, McCann M, Wilson C (eds) Psychology, Law and Criminal Justice. International Developments in Research and Practice, Waller de Gruyter, Berlin, pp 181–187

Xu F, Lee K (2007) Promoting honesty. Presented at the Society for Research in Child Development Biennial Meeting, Boston, MA

Xu F, Bao X, Talwar V, Lee K (under review) Prosocial lying in children: from concept to action. Dev Psychol

Zelazo PD, Muller U (2002). Executive function in typical and atypical development. In: Goswami U (ed) Blackwell handbook of childhood cognitive development. Blackwell Publishers, Malden, MA, pp 445–470

Zelazo PD, Carter A, Reznick JS, Frye D (1997) Early development of executive function: a problem-solving framework. Rev Gen Psychol 1:1–29

9 Discovering Mind: Development of Mentalizing in Human Children

Shoji Itakura[1], Mako Okanda[2], and Yusuke Moriguchi[3]

1. Introduction

For human infants, agents—other humans—are the fundamental units of their social world. Agents are very special stimuli to infants. Researchers of object–person differentiation have proposed a set of rules that infants may use during their interaction with people as opposed to objects. For example, Premack (1990) suggested that infants may perceive people as perceptual events that are both self-propelled and goal-directed objects. In such case, adults also perceive people as agents with intention. Spelke et al. (1995, p. 60) described an infant's concept of human as follows: "Three aspects of human interactions that are accessible in principle to young infants are contingency (humans react to one another), reciprocity (humans respond in kind to one to another's actions), and communication (humans supply one another with information)." Spelke et al. showed that infants may interpreter an object's movement with these three principles and the "principle of contact." To explain the contact principle, they used the habituation procedure and showed that infants tended to assume that an object, if it moves, should have been set in motion by the push from another object (or person). On the other hand, there is no need to apply an external force for a social agent to move. They demonstrated that this kind of perception of agency has appeared in 7-month-olds. Agents are not simply physical objects with new properties added to them. On the contrary, they are entities of an animacy that can move on their own, breath, eat, drink, look, and engage in actions with objects or interact with other agents (Gomez 2004).

From the point of view of social cognitive development, Johnson raised two questions (2003): (1) when do children first attribute mental state to others? and (2) when they do so, to whom do they attribute mental state?

[1] Department of Psychology, Graduate School of Letters, Kyoto University, Yoshida-Honmachi, Sakyo-ku, Kyoto 606-8501, Japan
[2] Research Fellow of the Japan Society for the Promotion of Science; Kyoto University, Yoshida-Honmachi, Sakyo-ku, Kyoto 606-8501, Japan
[3] Research Fellow of the Japan Society for the Promotion of Science; University of Tokyo, 3-8-1 Komaba, Meguro-ku, Tokyo 153-8902, Japan

In this chapter, we review a line of research that we have conducted to investigate how children understand and detect human agents and nonhuman agents. We first start with the cue that infants use to infer an agent as a social partner. The use of this type of cue is the ability to detect whether caretakers and social partners are attentive and responsive to their own behavior in social exchanges. We call this social contingency. We introduce two studies that we have conducted on infants' sensitivity toward the social contingency of their mothers and a stranger. In the second part, we describe a study investigating the relationship between understanding the other's mental states and applying their own inhibitory control in young children with card sorting tasks. In the third part, we describe two studies investigating an infant's interpretation of nonhuman agent action based on the work of Csibra and colleagues (Csibra 2003; Kuhlmeier et al. 2003).

Finally, we introduce a study on infants' imitation of a robot's action and propose a new research domain that we call "developmental cybernetics." Developmental cybernetics is a study on interaction between children and robots (Kojima 2005). It has been predicted that in common households in the 21st century robotics technology will be used as widely as refrigerators and dishwashers in today's family (Asada and Kuniyoshi 2006; Ishiguro and Miyashita 2005). Therefore, it is important to explore developmental cybernetics.

2. Sensitivity to Social Contingency in Infants

Sensitivity to social contingency is the first step of understanding people as social agents. Social contingency is a useful cue for an infant to distinguish his/her self from others. During social interaction, an infant may recognize that people are interactive if he/she reacts to them contingently; therefore, sensitivity to social contingency is an important milestone during the development of social cognition.

For the past 20 years, researchers have been interested in when infants begin to show their sensitivity to social contingency and how infants recognize a social partner's noncontingency. To investigate these questions, the still face paradigm (Striano 2004; Tronick and Brazelton 1978) and the double video (DV) live-replay paradigm were developed and have been widely used (Murray and Trevarthen 1985; Nadel and Reserbat-Plantey 1999). In the still face paradigm, a mother is asked to be unresponsive (i.e., to hold a still face) after she has engaged in a normal face-to-face interaction with her infant. It has been found that the infant tends to respond negatively upon seeing the mother's still face. With this paradigm, Adamson and Frick (2003) reported that infants' sensitivity to social contingency appears between 2 and 8 months.

Murray and Trevarthen (1985) invented the DV live-replay paradigm in which the mother and her infant first interact via monitors and cameras (live condition), and then the infant is presented with a replay of the recorded mother as a noncontingent episode (replay condition). In this paradigm, an infant is able to keep watching his/her mother who is interacting with him/her even during the noncontingency period, so their interaction is always two way. With this paradigm,

Murray and Treverthen (1985) found that 2-month-olds were sensitive to social contingency of their mothers.

However, some researchers argued that the observed behavioral changes (e.g., gaze and smile reduction) during the replay condition in infants may be the result of increased time instead of the infants' sensitivity to social contingency (Rochat et al. 1998). To clarify this issue, Nadel et al. (1999) used three uninterrupted sessions: Live 1, Replay, and Live 2. They found a complex V-shape pattern in terms of 2-month-olds' reaction: during Replay, there is a decrease of positive indices (gaze and smile at mother), but an increase of negative index (frowning), and during Live 2, such behavioral changes disappeared. This finding supports the statement that 2-month-olds have developed social contingency.

In addition, Nadel and colleagues (1999) used a seamless edit technique between different conditions, instead of including an external interrupt such as presenting the infant with a hand puppet (Rochat et al. 1998) or a black screen (Bigelow and DeCoste 2003). Studies using such seamless shift techniques have shown that indeed 2-month-olds are sensitive to social contingency (Bigelow and DeCoste 2003; Nadel et al. 1999). Nevertheless, it is still unknown whether infants younger than 2 months old are sensitive to the mother's social contingency. Thus, we conducted a study to investigate this issue.

In addition, we are also interested in the question of when infants show sensitivity to the social contingency of a stranger. As the mother is the first social partner for an infant, the infant is able to discriminate his/her mother from others at a very early age. One may expect that an infant may develop the sensitivity to the contingency of his/her mother at a very early age as well. Previous work has shown that indeed infants respond to the social contingency of strangers differently from their response to that of their mothers. For example, Hains and Muir (1996) reported that 5-month-old infants were sensitive to a stranger's social contingency but not to their mothers.' Hains and Muir (1996) and Nadel et al. (1999) reasoned that this was because the mother's noncontingency was not important for infants at this age because they had already developed a stable relationship with their mothers. However, infants did not have such a good relationship with the stranger, and therefore they responded negatively to the stranger when the stranger was unresponsive. Nevertheless, it is still unknown whether infants younger than 5 months old are sensitive to the noncontingency of a stranger. The current study was planned to investigate this question. Thus, we conducted two studies to investigate the developmental changes of social contingency of mothers and strangers in infants using a double video paradigm (Okanda and and Itakura, 2008) (Fig. 1).

Infants' behaviors during the first contingent (Live 1, 30 s), noncontingent (Replay, 30 s), and the second contingent (Live 2, 30 s) conditions were analyzed. Given the possibility that infants may not recognize a social partner's noncontingency immediately after replay begins, each condition was divided into two periods, the first being 15 s and the second 15 s.

As shown in Fig. 2, when the mother became noncontingent, younger infants showed an increase of gaze from the first 15 s to the second 15 s of the replay

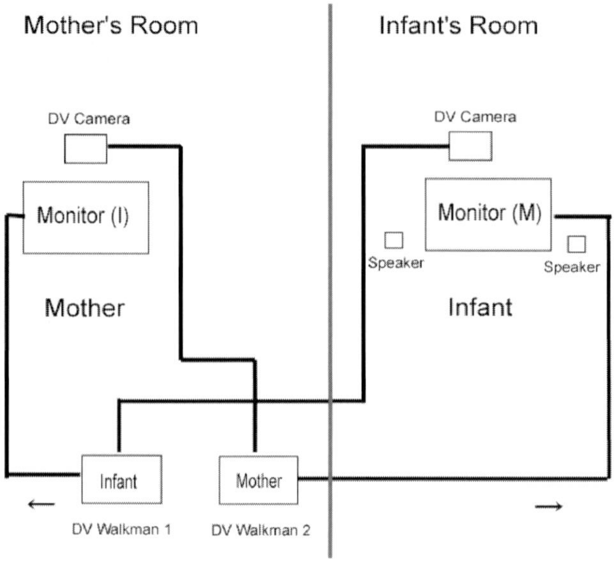

FIG. 1. Schematic representation of experimental situation.

condition. When the stranger became noncontingent, only older infants responded differently by showing an increase of smile during the replay condition.

In conclusion, these results showed that infants younger than 2 months old were able to detect the mother's noncontingency. In addition, the results suggest that infants may recognize the mother's change (noncontingency) very early. Thus, the current study provides a new piece of evidence of early social cognitive development. Moreover, the current study also showed that 4-month-olds were sensitive to a stranger's social contingency, and they emotionally reacted to stranger's noncontingency. They smiled to make the stranger contingent again, which was an active behavior. These results imply that there may be three components in sensitivity to social contingency: detection, response, and expectancy. Among these components, detection is the basic component to establish sensitivity. At an early age, these three components may be passive reactions but may develop into active responses as the child grows older. In addition, certain social tools may assist such transition as well.

3. Social Transmission of Disinhibition

Inhibitory control plays an important role in multiple areas of child development such as attention, thinking, communication, and social interaction. This is one of the most rapidly developing cognitive abilities in preschool years. Numerous studies have shown that between 3 and 5 years of age children develop inhibitory control rapidly (Carlson 2005; Diamond et al. 2002; Zelazo and Müller 2002).

One widely used inhibitory control task is the Dimensional Change Card Sorting task (DCCS; Zelazoet al. 1996). In this task, children are instructed to

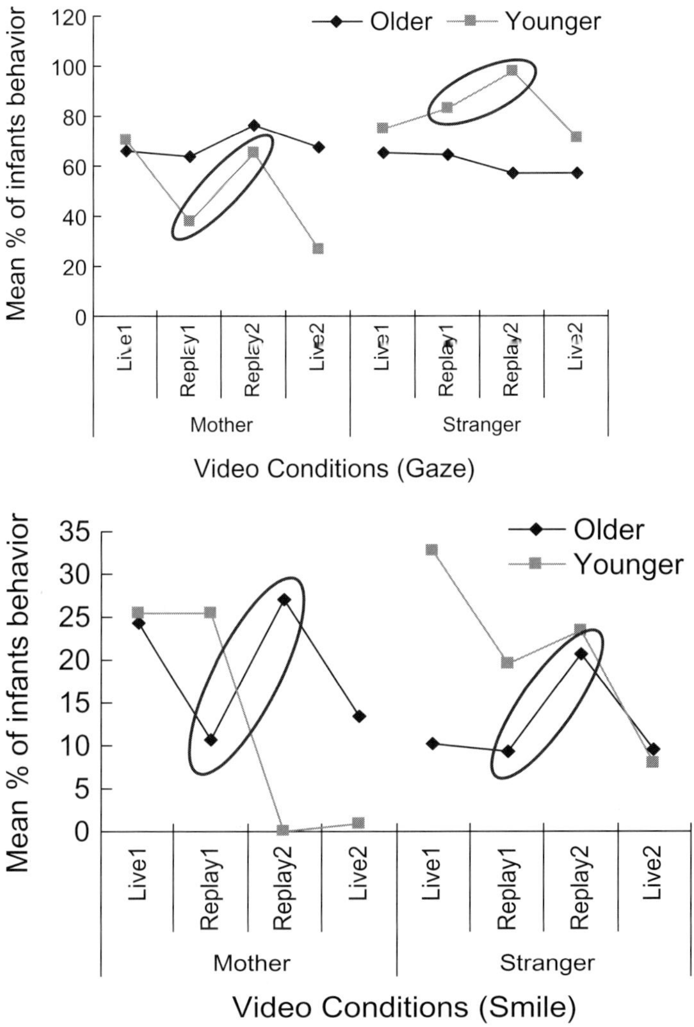

a

b

FIG. 2. **a** Mean percentage of gazing time **b** Mean percentage of smiling behavior.

sort cards that have two dimensions (e.g., color and shape). Children are first asked to sort cards by one dimension (e.g., color). Then, children are told to stop sorting cards by the previous dimension and to start to sort the cards according to the other dimension (e.g., shape).

Many studies have shown that older preschoolers are able to switch easily from one to the other dimension. However, most younger preschoolers have difficulty, after correctly sorting the cards according to the first rule in the first phase, in switching their sorting according to the second dimension in the second phase. They persevere and continue to sort the cards according to the first rule, in spite of the fact that they appeared to have no difficulty describing the rule change verbally (Zelazo et al. 1996). In addition, 3-year-olds who failed to switch to the new rule by themselves also insisted that a puppet should follow the old rule as well (Jaques et al. 1999). Three-year-olds' perseveration on the DCCS is robust (see review in Zelazo et al. 2003) and has been observed even in Asian countries and cultures (Sabbagh et al. 2006). However, it should be noted that some procedural modifications may alleviate young children's difficulty. For example, Munakata and Yeris (2001) found that modification of the instructions of the two rules led to children's overcoming the perseverative errors. Towse et al. (2000) implied that an experimenter's demonstration of sorting according the second rule in the second phase may reduce children's tendency to perseverate.

Compared with the large amount of research in the development of inhibitory control and the emphasis of social environment on development in general, the impact of social factors on 3-year-olds' perseverative errors has not been systematically investigated, in spite of the fact that many recent studies have shown significant correlations between children's inhibitory control abilities and their social cognition in various domains such as emotional knowledge, moral conscience, and theory of mind understanding (Carlson et al. 2002; Eisenberg et al. 1995; Hughes 1998; Hughes et al. 1998; Kochanska et al. 1996, 2000; Perner et al. 2002). Although the existing evidence regarding the linkage is mainly correlational, many researchers have speculated that inhibitory control abilities could provide a crucial foundation for the development of social cognition (Sabbagh et al. 2006; Zelazo et al. 2005; Zelazo and Müller 2003). However, it is equally possible that social cognition may contribute to improvement in children's inhibitory control abilities (Perner 1998). Indeed, Kloo and Perner (2003) showed that training children on theory of mind tasks leads to improvement in their performance on the DCCS task (although training on DCCS also improves children's performance on theory of mind tasks).

To illustrate that social cognition can influence inhibitory control, Moriguchi et al. (2007) conducted a series of experiments. They hypothesized that social information should have an effect on children's inclination to commit perseverative errors in the DCCS. A modified DCCS task was used. In Experiment 1, during pre-switch, instead of sorting the cards by themselves, children observed an adult model sorting the cards according to one rule (e.g., the shape rule). Then, during post-switch, children were asked to sort according to a different rule (e.g., the color rule). They examined whether children would show perseveration and fail to sort the cards according to the second rule after merely observing the adult model sorting the cards according to the first rule.

Figure 3 depicts the correct response distributions of children in each age. As shown in this figure, 68% of the 3-year-old children failed to follow the new rule at least on four trials. They rather continued to follow the rule which the demonstrator employed to sort the cards. In contrast, 35% of the 4-year-old and 10% of the 5-year-old children showed perseverative errors. These results are consistent with the results of previous study employing the DCCS task. Observation appeared to influence younger children's performance. They appeared to have difficulty in inhibiting the old rule, although employed by the other person, and

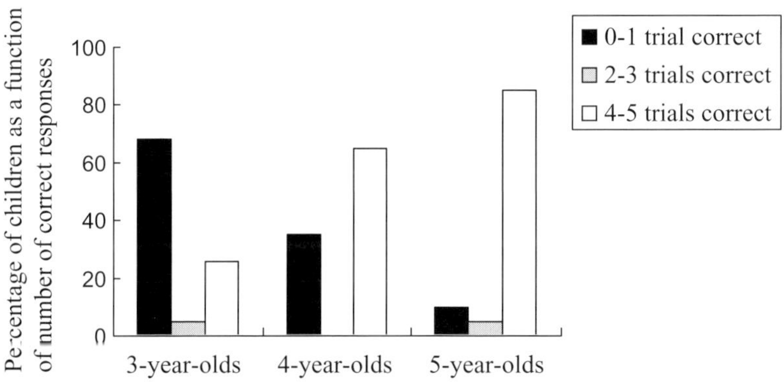

FIG. 3. Correct response distributions of children at each age.

switching to use a new rule to sort cards by themselves. Observation of the other person's behavior seems to have a "contagion effect" on younger children. This result showed that, in young children, perseveration can occur through social transmission from one person to another.

In Experiment 2, the possibility that the children might have merely preferred to imitate another person's motor behavior rather than to follow the experimenter's instruction was tested. In this experiment, 3- and 4-year-old children first observed an adult model sorting cards in a wrong way. In one condition, the model seemed to be aware that she was not following the experimenter's instruction but continued to sort the cards incorrectly anyway (Aware condition). In the other condition, the adult model seemed not to be aware that she was sorting the cards incorrectly and thus continued to sort the cards based on an incorrect rule (Unaware condition). After observing the model's sorting, the subjects were instructed to sort the cards according to the same rule as the one that the model was instructed to follow (Fig. 4).

Children performed significantly more poorly in the Unaware condition than in the Aware condition. In the Unaware condition, 78% of the 3-year-old and 42% of the 4-year-old children showed perseverative errors. In contrast, in the Aware condition, even younger children did not show perseverative errors. These results suggest that children's card sorting behaviors were mostly affected by the information provided by the model, not by the experimenter. However, there might be another explanation for the results of this experiment. There is a possibility that the experimenter's feedback given to the adult model might have different effects on children's card sorting in each condition. For example, in the Aware condition, the experimenter gave feedback of agreement to the model when she became aware of her mistakes. On the other hand, in the Unaware condition, the adult model was not aware of her mistakes, but the experimenter pointed out her mistakes. These differences of feedback by the experimenter may make the children confused. So it was possible that children's different

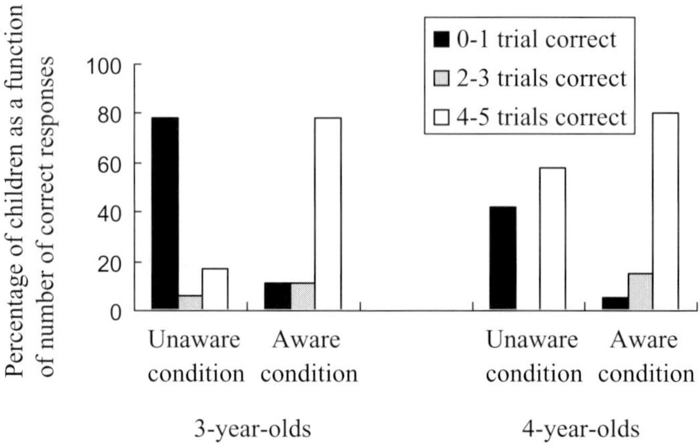

FIG. 4. Correct response distributions of children at each age.

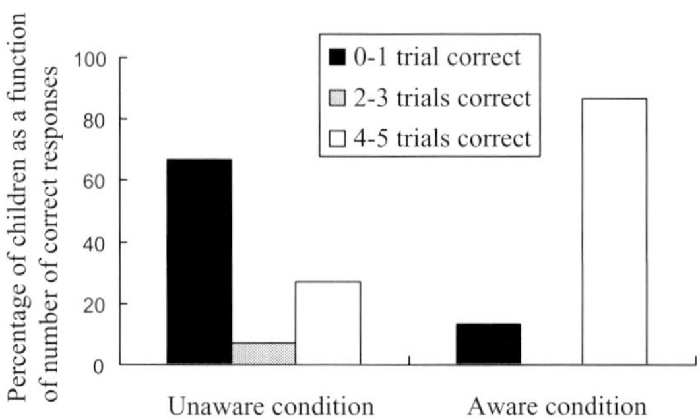

FIG. 5. Correct response distribution in each condition.

responses were caused by the experimenter's feedback. To test this possibility, Moriguchi et al. (2007) conducted Experiment 3 with 3-year-old children.

In Experiment 3, the procedure was same as in Experiment 2 except that the experimenter did not give any feedback to the adult model during the observation phase (Fig. 5). This figure depicts correct response distribution of 3-year-old children in Unaware condition and Aware condition.

Although in the Unaware condition 67% of the 3-year-old children made perseverative errors, in the Aware condition, 13% of the 3-year-old children showed such errors. Without the experimenter's feedback to the adult model, 3-year-old children in the Unaware condition continued to commit perseverative errors. These results may suggest that the model's statement about her own knowledge regarding whether she had followed the correct rule is crucial for 3-year-old children's performance in understanding instead of the experimenter's feedback.

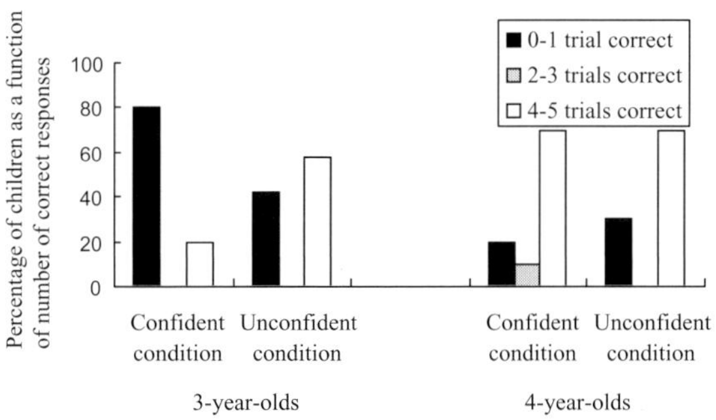

FIG. 6. Correct response distributions of children at each age.

In the final experiment, Experiment 4, we tested whether the degree of the model's confidence about card sorting influences the children's perseverative errors in card sorting task in a similar situation to Experiment 1. The procedure of the situation was identical to that of Experiment 1. There were two conditions, Confident condition and Unconfident condition. In the Confident condition, the model expressed confidence in her sorting, whereas in Unconfident condition the model expressed concern that she may not have followed the experimenter's instruction correctly (Fig. 6).

Most 3-year-old children in the Confident condition consistently showed perseverative errors. Nevertheless, in the Unconfident condition more than half the 3-year-old children successfully switched to the second sorting rule. In contrast to 3-year-olds, 4-year-olds performed similarly in these conditions. These results suggest that the social pragmatic information of the model's confidence affected 3-year-old children but not 4-year-olds, possibly because most 4-year-old children have no difficulty inhibiting the previous rule and executing sorting based on a new rule.

These findings suggest that perseveration can be transmitted not only from one task to another in the same child, but also socially from one person to another. We call such transmission "social contagion."

4. Interpretation of Intentional Mental States in Geometric Figures by Infants

People have a robust inclination to perceive meaningful social causality. This kind of inclination is so pervasive that we are keen to attribute goal-directedness, intentions, or mental state, even personality, to geometric figures based on the way they move on a screen (Rochat 2003). Studies on *perceptual social illusion* have demonstrated that dispositional qualities and intention can be systematically perceived in the context of special sequential movements of two or more geometric figures (Heider and Simmer 1944; Basili 1976; Dittrich and Lea 1994).

According to Csibra and his colleagues, there are two ways to attribute intentional mental states to others upon looking at their actions (Csibra 2003). One way is teleological stance, which is that people interpret an observed behavior as a *goal-directed action*. A goal-directed action is "about" the end state of that action; it is performed to make the end state occur (Csibra 2003). Using an ingenious violation-expectation paradigm, Csibra and his colleagues demonstrated that 12-month-olds understand goal-directed actions. In their experiment, the infants were first habituated to a computer-animated goal-directed event in which a small disk tried to reach and contact a large disk by jumping over an obstacle (Fig. 7). In this case, the small disk could be perceived as a social agent and the large disk could be its "goal." Then, during the test phase, the obstacle was removed, and infants were shown either the original jumping goal-approach or a perceptually new straight-line goal-approach. Results showed that 12-month-olds looked longer at the original jumping action, but indicated no dishabituation to the novel straight-line action. Csibra et al. suggested that 12-month-olds looking longer at the old action may be caused by violation-of-expectation. Because there was no obstacle, jumping seemed to be an inefficient means to the goal now. In contrast, the new straight-line action appeared to be the most efficient means to the goal given the fact that there was no obstacle situation. Thus, infants did not show any dishabituation to the new action. These results suggest that by 12 months infants can (1) interpret others' actions as goal directed, (2) evaluate which one of the alternative actions available within the constraints of the situation is the most efficient means to the goal, and (3) expect the agent to perform the task by the most efficient means available (Gergely and Csibra 2004). Therefore, 12-month-olds appear to have the ability of teleological inference about goal-directed actions.

In addition to understand others' goal-directed actions, infants at this age can also predict others' future actions by observing their pre-events. Kuhlmeier et al. (2003) demonstrated that 12-month-old infants could recognize goal-directed

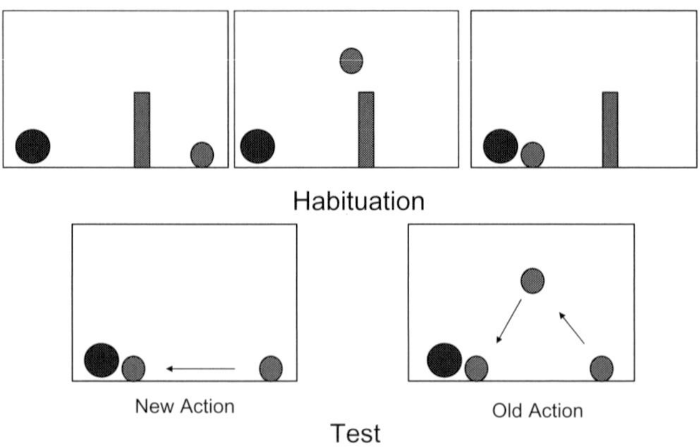

Fig. 7. Stimuli used in experiments by Csibra et al.

action of animated geometric figures and infer their future actions. In their experiment, infants were habituated to an animated figure movie. In the movie, one object (the circle in this case) attempted to climb a hill; the other two objects (a square and a triangle) engaged in either a helping action or a hindering action toward it (see Fig. 8). After habituation, the infants were then shown two movies in which all three objects (a square, triangle, and circle) were presented in a new context. In one movie, the circle approached and settled next to its helper, and in the other movie, the circle approached and settled next to its hinderer. Kuhlmeier et al. (2003) showed that 12-month-old infants tended to look longer at the movie depicting that the circle approached the helper than the movie in which the circle approached the hinderer. These results implied that 12-month-old infants could discriminate the helper and the hinderer with respect to the circle's new goal in the new context, and that the infants found the act of approaching the helper more coherent than the act of approaching the hinderer, possibly because they have posited a mentalistic mediator for the circle's actions.

In addition to replicating the study by Kuhlmeier and colleagues (2003), we added a "no-hill" condition in which the ball rolled along a flat surface (Tsuji and Itakura 2003). There were two experiments: the first aimed to replicate the

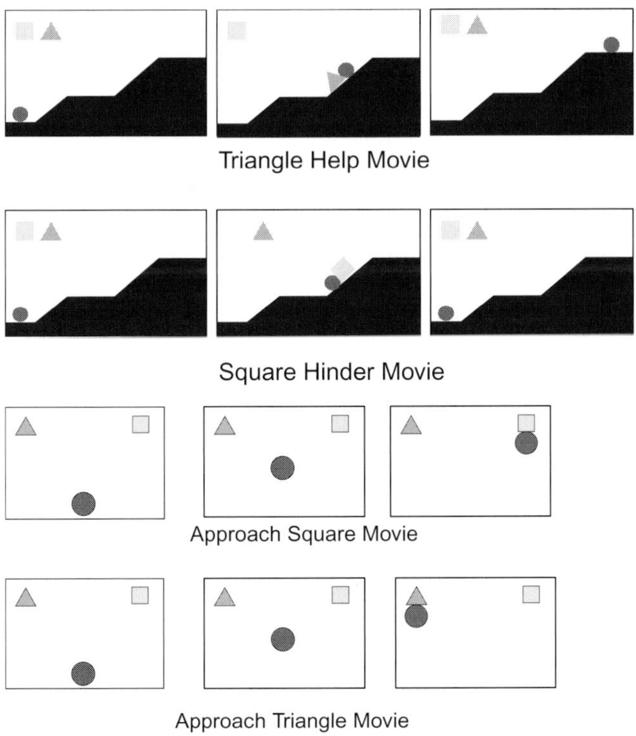

FIG. 8. Stimuli used in the experiments.

Kuhlmeier et al. experiment and the second was identical except for the addition of the no-hill condition. In Experiment 1, participants were 12-month-old infants. The procedure was similar to Kuhlmerier et al.'s study. An example of the stimuli used is provided in Fig. 8.

Experiment 1 replicated Kuhlmeier et al's (2003) findings by showing that infants looked longer at the movie in which the circle approached the helper.

In Experiment 2, the participants were 12-month-old infants. The procedure was almost the same as Experiment 1 except for the habituation stimuli. In Experiment 1, the circle climbed a hill; however, there was no hill in Experiment 2. We excluded the hill because we hypothesized that the existence of a hill emphasized the goal of the circle. In this condition, only the track of circle movement was the cue of its goal-like behavior. We predicted that infants would show no difference in looking time between the conditions in which the stimulus circle approaches the helper versus the hinderer. However, in contrast to our hypothesis, the results were the same as in Experiment 1. During the test trials, infants showed preferential looking toward the ball approaching the helper even when there was no hill present during the previous habituation.

It is still a puzzle how one approach event might be judged as preferable to the other (Kuhlmeier et al. 2003). Kuhlmeier et al. (2003) claimed that in their pilot study adult participants tended to see the circle as "liking" or "preferring" the helper object, because the circle completed its goal during the habituation period. However, it is unclear why the infants looked longer at the event when the circle approached the helping object, rather than showing surprise at the event when the circle approached the hindering object. Although these results are informative, questions remain. A more sophisticated and ecologically valid paradigm is needed to answer these questions.

5. Developmental Cybernetics

In the previous section, we discussed infants' interpretation of computer-animated figures. In this section, we introduce a new exciting research field on the development of mentalizing called *developmental cybernetics*. Future robots will not only perform household chores but may also serve as caregivers and educators to children. To date, there is no scientific evidence to ascertain whether children, particularly younger ones, will be amenable to receive care, let alone learn, from robots as readily as they do from humans. Despite recent rapid growth in research on developmental cybernetics, it is entirely unknown as to what essential human characteristics must be built into a robot to facilitate such learning.

We focused on our original studies relative to these issues.

One of the earliest fundamental forms of learning from another human is imitation. Imitation begins at birth with neonates imitating adult behaviors that are within their innately endowed behavioral repertoire (e.g., tongue protrusion: Meltzoff and Moore 1977, 1997; for a review, see Meltzoff 2005). With increased age, infants begin to imitate novel behaviors performed by adults, either live or televised (Barr and Hayne 2000; see Meltzoff 2005). Also, they are able to reenact an adult's novel behavior even after a long delay (Bauer et al. 2001). More strikingly, Meltzoff (1995) demonstrated that when adults performed an action that appeared to fail to accomplish their intended goal (e.g., instead of pulling apart two halves of a dumbbell, the adults' hands slipped and the dumbbell stayed intact), 18-month-olds were able to "imitate" the unobserved but intended act (e.g., pulling apart the dumbbell) rather than the observed but unintended act (e.g., the slippage of the adult's hands off the ends of the dumbbell, with the two halves not separated).

Similar results have been found in 15 month olds (Bellagamba and Tomasello 1999; Johnson et al. 2001; Meltzoff 1999) but not in 12-month-olds (Bellagamba and Tomasello 1999). Similarly, Carpenter et al. (1998) showed that 14- to 18-month-olds were more inclined to imitate an adult's intended actions than accidental actions. These findings indicate that by the second year of life, infants do not blindly imitate the behavior of others, but rather base imitation on their understanding of the intentions and goals of others. This development is perhaps built on another developmental milestone at around 9 to 12 months of age when infants begin to understand adults' behavior to be goal directed and intentional (Baldwin et al. 2001; Behne et al. 2005; Luo and Baillargeon 2005; Phillips and Wellman, 2005; Shimizu and Johnson 2004; Woodward 1998; see Gergely and Csibra 2004).

Why are human infants so inclined to imitate another person's behavior to the extent that they even "imitate" the person's intended but unconsummated acts? Meltzoff (2005, 2007) proposed a "Like Me" hypothesis. The central tenet of this hypothesis is that infants are innately endowed with the ability to see correspondence between the actions of others and those performed by their own body. With experience, infants learn to map between their own actions and failed actions with their internal mental states. This innate capacity to construe others' actions as "me relevant" coupled with an acquired understanding of their own mental state allows infants to crack the problem of the other minds. They use their own intentional actions as a framework for interpreting the intentional actions of others. As a result, they can imitate selectively another's intended, but not their unintended, actions. The existing developmental evidence of infant imitation involving humans as models is largely consistent with the "Like Me" hypothesis. It is also consistent with evidence that infants did not produce the target act when a mechanical device's behavior failed to complete an action (e.g., pulling the dumbbell apart; Meltzoff 1995). This inanimate device did not look at the human or interact with the target in a human fashion, either of which might have been sufficient not to trigger the "Like Me" interpretive framework.

What are the basic characteristics of an agent that enable infants to make "me relevant" mapping, infer the agent's goals, and thus imitate the agent's intended but unconsummated actions? One possibility is that such an agent must share human morphological characteristics (MacDorman and Ishiguro 2006a,b). This suggestion seems reasonable given the evidence that person recognition in general and face recognition in particular begin in early infancy and develop rapidly (Johnson and Morton 1991; Quinn and Slater 2003). An ability to recognize and interpret faces can in theory serve as an essential enabling factor for infants to carry out the "like me" mapping and thus successful imitation of intentional and goal-directed actions.

However, Johnson and her colleagues (Johnson et al. 1998; Johnson 2003) suggested that infants' recognition of intentional agents is not necessarily isomorphic with person recognition but rather is based on a set of nonarbitrary object recognition cues. Johnson et al. (2001) showed that a novel orangutan-like object (with eyes and nose but no mouth) that appeared to be self-propelled and interacted with infants contingently led 15-month-olds to imitate its unconsummated acts. Further, infants also displayed significantly more communicative behaviors toward the orangutan-like object than toward another physically similar but faceless and inanimate object.

The Johnson et al. (2001) results clearly indicate that full-fledged human morphological characteristics are not necessary to engender imitation of intentional acts in infants. However, it was unclear whether infants' imitation of intentional acts was engendered by behavioral similarities between humans and the orangutan-like object (e.g., self-propelled and contingent movements) or the morphological similarities between the two (particularly their eyes). For example, infants might have attributed intentions to the object because of its eyes. Indeed, infants at birth are already sensitive to stimuli containing eyes (Farroni et al. 2004). With increasing age, they increasingly treat objects with eyes substantially different from those without eyes (Graham et al. 2007). Furthermore, before they can perform imitation of intended but unconsummated acts, they already are able to use another's eye gazes to infer their mental states (Brooks and Meltzoff 2002). Thus, it is possible that the presence of a pair of eyes alone is sufficient for infants to "imitate" an agent's intentional acts. Alternatively, the presence of eyes must be coupled with certain contingent and meaningful actions to ensure imitation of intentional acts. The present study tested these possibilities.

Itakura et al. (in press) modified Meltzoff's (1995) paradigm. Instead of human adults, a robot named Robovie, with eyes and mechanical arms, served as a model (Fig. 9). Robovie was developed at the ATR Intelligence Robotics and Communication Laboratory in Japan.

Similar to human models observed by Meltzoff (1995), the robot performed novel actions either successfully or unsuccessfully. The robot's behavior was videotaped and presented to children between 24 and 35 months of age on a television monitor. We chose this age range because existing studies (Bellagamba and Tomasello 1999; Johnson et al. 2001; Meltzoff 1995) have shown that by 2 years of age children can successfully imitate an adult's intended but unconsummated actions. In the Eye Contact Condition, both before and after performing a novel action the robot made eye contact with a human adult who was also present throughout the video presentation. In the No Eye Contact Condition, although the human adult was present and behaved exactly the same way as in the Eye Contact Condition, the robot did not make eye contact. Thus, in the Eye

FIG. 9. Humanoid robot "Robovie." (Photograph from ATR Intelligence Robotics and Communication Laboratory).

Contact and No Eye Contact conditions, eyes were present. If the presence of eyes alone is sufficient, children should correctly imitate both the successful and unsuccessful acts performed by the robot in both conditions. Otherwise, children should succeed in imitating the unconsummated acts in the Eye Contact Condition but not in the No Eye Contact Condition.

Three sets of objects, modeled after Meltzoff (1995), were used: a dumbbell, a cup and beads, and a peg with an elastic band. Robovie was controlled to types of action trials. In the Successful Demonstration condition, Robovie pulled the dumbbell apart, put the beads into the cup, and hung the elastic band on the peg. In the Unsuccessful Demonstration condition, Robovie tried to pull the dumbbell apart, but the dumbbell remained intact. The robot tried to put the beads into the cup, but the beads dropped outside the cup. The robot also tried to hang the elastic band on the peg, but the elastic band fell on the table.

Children were divided into five conditions: Successful Demonstration + Eye Contact Condition, Successful Demonstration + No Eye Contact Condition, Unsuccessful Demonstration + Eye Contact Condition, Unsuccessful Demonstration + No Eye Contact Condition, and a baseline condition in which children were simply given one of the objects to manipulate. Children were coded as having produced the target action if they showed such behavior (Fig. 10).

Results in the Successful Demonstration Condition showed that young children imitated successful actions regardless of whether the robot made eye contact

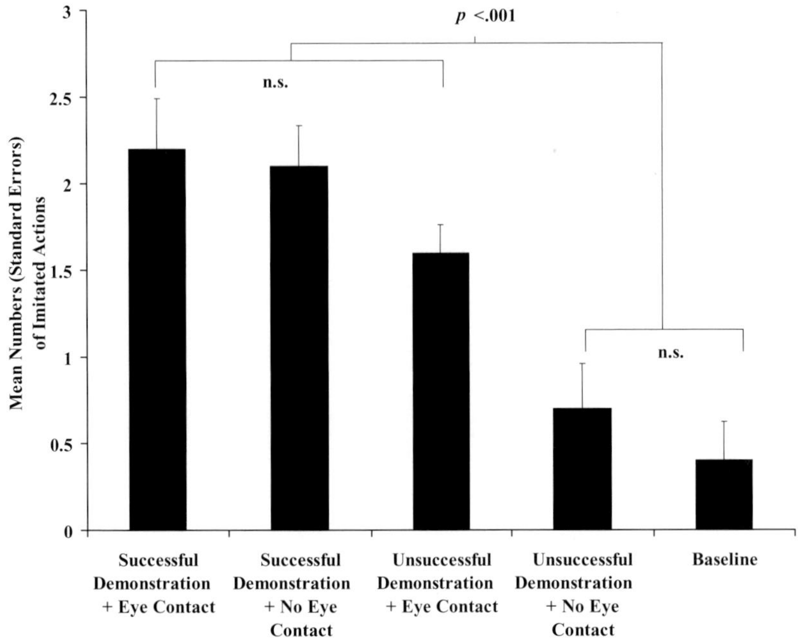

FIG. 10. Mean (standard error) number of imitated actions for each condition.

with a human. In the Unsuccessful Demonstration Condition, however, children completed the unobserved but intended action only when the robot made eye contact with the human.

There are two main findings in this study. First, young children imitate a nonhuman agent's action. Second, eye contact must be coupled with interactive activities with another human to complete the action of intentional actions. These findings will help robotic scientists design robots that not only mimic human morphology and biomechanical movements but also convey a sense of "intentionality."

6. Conclusions

In this chapter, we focused on the development of mentalizing in human children. Sensitivity to social contingency is the first demonstration of the ability for mentalizing in infancy. Even 1-month-old infants can detect unusual responses of the social partner, but this awareness is limited to only their mothers. By age of 4 months, infants become more sensitive to a stranger's contingency. In addition, 4-month-old infants use a smile as a social tool when the mother's responses are noncontingent. From early infancy, infants are able to perceive their social partner as an agent just like themselves.

Children's behaviors tend to be affected easily by other social agents. Young children's performance on the DCCS task can be influenced by their observation of how others perform the task. Interestingly, younger children can also consider the mental states of the adult model when deciding whether to use the old rule or the new rule. When the adult model appeared uncertain whether she sorted correctly, most 3-year-olds no longer committed perseverative errors. The False Belief task has been used widely to measure mentalization in preschoolers. Our results suggest that social transmission can be another useful tool to reveal the development of theory of mind in young children.

Then how does such ability develop? What is the basis of this kind of ability? When do human infants start to explain and predict other's actions by attributing causal intentional mental state to them? Recent studies demonstrated that 1-year-old infants interpret and draw inferences about others' goal-directed actions. Gergely and Csibra (2003) claimed that this early inferential principle is a representational system that develops later to guide inferences about mental states. Around 15 months of age, infants can infer disposition of the animated stimuli. Reliable imitation of intentional actions performed by others emerges at approximately at 18 months. All these abilities lead to the development of theory of mind later. We proposed "developmental cybernetics" as a completely new field to investigate the development of mentalization in children. We used a nonhuman agent such as a robot as a tool to clarify under what conditions children can attribute goal-directedness, intention, and mental states to others. We found that subtle actions added to the robot, such as eye contact and gazing, are very effective to change the infants' reenactment behavior. Developmental cybernetics should be useful both in investigating the development of social cognition in children and in designing robots for human life.

References

Adamson LB, Frick JE (2003) The still face: a history of a shared experimental paradigm. Infancy 4:451–473

Asada M, Kuniyoshi Y (2006) Robot intelligence, vol 4. Iwanami Shoten, Tokyo

Baldwin DA, Baird JA, Saylor MM, Clark MA (2001) Infants parse dynamic action. Child Dev 72:708–717

Barr R, Hayne H (2000) Age-related changes in imitation: implications for memory development. In: Rovee-Collier C, Lipsitt LP, Hayne H (eds) Progress in infancy research. Erlbaum, Hillsdale, NJ, pp 21–67

Basili JN (1976) Temporal and spatial contingencies in the perception of social events. J Pers Soc Psychol 33:680–685

Bauer PJ, Wiebe SA, Waters JM, Bangston SK (2001) Re-exposure breeds recall: effects of experience on 9-month-olds' ordered recall. J Exp Child Psychol 80:174–200

Behne T, Carpenter M, Call J, Tomasello M (2005) Unwilling versus unable: infants' understanding of intentional action. Dev Psychol 41:328–337

Bellagamba F, Tomasello M (1999) Re-enacting intended acts: comparing 12- and 18-month-olds. Infant Behav Dev 22:277–282

Bigelow AE, DeCoste C (2003) Sensitivity to social contingency from mothers and strangers in 2-, 4-, and 6-month-old infants. Infancy 4:111–140

Blair C (2002) School readiness: integrating cognition and emotion in a neurobiological conceptualization of children's functioning at school entry. Am Psychol 57:111–127

Brooks PJ, Hanauer JB, Padowska B, Rosman H (2003) The role of selective attention in preschoolers' rule use in a novel dimensional card sort. Cognit Dev 18:195–210

Brooks R, Meltzoff AN (2002) The importance of eyes: how infants interpret adult looking behavior. Dev Psychol 38:958–966

Carlson SM (2005) Developmentally sensitive measures of executive function in preschool children. Dev Neuropsychol 28:595–616

Carlson SM, Moses LJ, Breton C (2002) How specific is the relation between executive function and theory of mind? Contributions of inhibitory control and working memory. Infant Child Dev 11:73–92

Cowley SJ, Kanda T (2005) Friendly machines: interaction oriented robots today and tomorrow. Alternation 12:79–106

Carpenter M, Nagell K, Tomasello M (1998) Social cognition, joint attention, and communicative competence from 9 to 15 months of age. Monogr Soc Res Child Dev 63(4): serial no. 255

Csibra G (2003) Teleological and referential understanding of action in infancy. Philos Trans R Soc Lond Ser B Biol Sci 358:447–458

Dittrich WH, Lea SEG (1994) Visual perception of intentional motion. Perception 23:253–268

Eisenberg N, Fabes RA, Murphy M, Maszk P, Smith M, Karbon M (1995) The role of emotionality and regulation in children's social functioning. Child Dev 66: 1239–1261

Farroni T, Massaccesi S, Pividori D, Johnson MH (2004) Gaze following in newborns. Infancy 5:39–60

Gergely G, Csibra G (2004a) Teleological reasoning in infancy: the naïve theory of rational action. Trends Cognit Sci 7:287–292

Gergely G, Csibra G (2004b) The social construction of the cultural mind: Imitative learning as a mechanism of human pedagogy. Interaction Studies: Social Behaviour and Communication in Biological and Artificial Systems. Special Issue: Making Minds II(6):463–481

Gergely G, Nadasdy Z, Csibra G, Biro S (1995) Taking the intentional stance at 12 months of age. Cognition 56:165–193

Gomez JC (2004) Apes, monkeys, children, and the growth of mind. Harvard University Press, Harvard

Graham SA, Nilsen ES, Nayer SL (2007) Following the intentional eye: the role of gaze cues in early word learning. In: Flom R, Lee K, Muir D (eds) Ontogeny of gaze processing. Erlbaum, Mahwah, NJ

Hains SMJ, Muir DW (1996) Effect of stimulus contingency in infant-adult interactions. Infant Behav Dev 19:49–61

Heider F, Simmer S (1944). An experimental study of apparent behavior. Am J Psychol 57:243–259

Hughes C (1998) Executive function in preschoolers: links with theory of mind and verbal ability. Br J Dev Psychol 16:233–253

Hughes C, Dunn J, White A (1998). Trick or treat? Uneven understanding of mind and emotion and executive dysfunction in "hard to manage" preschoolers. J Child Psychol Psychiatry 39:981–994

Ishiguro H, Miyashita T (2005) Communication robot: the technology for development of the robots interacting with humans. Ohmsha, Tokyo

Itakura S, Ishida H, Kanda T, Lee K, Ishiguro H (in press) How to build an intentional android: infants' imitation of a robot's goal-directed actions. Infancy
Johnson MH, Morton J (1991) Biology and cognitive development: the case of face recognition. Blackwell, Oxford
Johnson SC (2003) Detecting agents. Philos Trans R Soc Lond Ser B Biol Sci 358:549–559
Johnson SC, Booth A, O'Hearn K (2001). Inferring the goals of nonhuman agent. Cognit Dev 16:637–656
Johnson SC, Slaughter V, Carey S (1998) Whose gaze will infants follow? Features that elicit gaze-following in 12-month-olds. Dev Sci 1:233–238
Kloo D, Perner J (2003) Training transfer between card sorting and false belief understanding: helping children apply conflicting descriptions. Child Dev 74:1823–1839
Kloo D, Perner J (2005) Disentangling dimensions in the dimensional change card-sorting task. Dev Sci 8:44–56
Kochanska G, Murray KT, Jacques TY, Koenig AL, Vandegeest KA (1996) Inhibitory control in young children and its role in emerging internalization. Child Dev 67: 490–507
Kochanska G, Murray KT, Harlan ET (2000) Effortful control in early childhood: Continuity and change, antecedents, and implications for social development. Dev Psychol 36:220–232
Kojima H (2005) Children–robot interaction: from interaction to cognition. Presented at the ESF research conference on brain development and cognition in human infants, "From action to cognition," Acquafredda di Maratea, Italy
Kuhlmeier VWK, Bloom P (2003) Attribution of dispositional states by 12-month-olds. Psychol Sci 14:402–408
Luo Y, Baillargeon R (2005) Can a self-propelled box have a goal? Psychological reasoning in 5-month-old infants. Psychol Sci 16:601–608
MacDorman KF, Ishiguro H (2006a) The uncanny advantage of using androids in social and cognitive science research. Interact Stud 7(3):297–337
MacDorman KF, Ishiguro H (2006b) Opening Pandora's uncanny box: Reply to commentaries on "The uncanny advantage of using androids in social and cognitive science research." Interact Stud 7(3):361–368
Meltzoff AN (1995) Understanding the intentions of others: Re-enactment of intended acts. Dev Psychol 31:838–850
Meltzoff AN (1999) Origins of theory of mind, cognition and communication. J Commun Disord 32:251–269
Meltzoff AN (2005) Imitation and other minds: the "Like Me" hypothesis. In: Hurley S, Cater N (eds) Perspectives on imitation: from neuroscience to social science, vol 2. MIT Press, Cambridge, pp 55–77
Meltzoff AN (2007) The "Like Me" framework for recognizing and becoming an intentional agent. Acta Psychol 124:26–43
Meltzoff AN, Moore MK (1977) Imitation of facial and manual gestures by human neonates. Science 198:75–78
Meltzoff AN, Moore MK (1997) Explaining facial imitation: a theoretical model. Early Dev Parent 6:179–192
Moriguchi Y, Lee K, Itakura S (2007) Social transmission of disinhibition in young children. Dev Sci 10:481–491
Muir DW, Nadel J (1998) Infant social perception. In: Slater A (ed) Perceptual development: visual, auditory, and speech perception in infancy. Psychology Press, London, pp 247–285

Munakata Y, Yeris BE (2001) All together now: when dissociations between knowledge and action disappear. Psychol Sci 12:335–337
Murray L, Trevarthen C (1985) Emotional regulation of interactions between two-month-olds and their mothers. In: Field TM, Fox NA (eds) Social perception in infants. Ablex, Norwood, NJ, pp 177–197
Nadel J, Carchon I, Kervella C, Marcelli D, Reserbat-Plantey D (1999) Expectancies for social contingency in 2-month-olds. Dev Sci 2:164–173
Okanda M, Itakura S (2008) One-month old infants' sensitivity to social contingency from mothers and strangers. Psychol Rep 102:293–298
Perner J (1998) The meta-intentional nature of executive function and theory of mind. In: Carruthers P, Boucher J (eds) Language and thought. Cambridge University Press, Cambridge
Perner J, Lang B, Kloo D (2002) Theory of mind and self-control: more than a common problem of inhibition. Child Dev 73:752–767
Phillips AT, Wellman HM (2005) Infants' understanding of object-directed action. Cognition 98(2):137–155
Poulin-Dubois D (1999) Infant's distinction between animate and inanimate objects: the origin of naïve psychology. In: Rochat P (ed) Early social cognition. Erlbaum, Hillsdale, NJ, pp 257–280
Premack D (1990) The infant's theory of mind. Cognition 36:1–16
Quinn PC, Slater A (2003) Face perception at birth and beyond. In: Pascalis O, Slater A (eds) The development of face processing in infancy and early childhood: current perspectives. Nova, Hauppauge, NY, pp 3–11
Rochat P, Neisser U, Marian V (1998) Are young infants sensitive to interpersonal contingency? Infant Behav Dev 21:355–366
Sabbagh M, Xu F, Carlson S, Moses L, Lee K (2006) The development of executive functioning and theory-of-mind: a comparison of Chinese and U.S. preschoolers. Psychol Sci 17:74–81
Shimizu YA, Johnson SC (2004) Infants' attribution of a goal to a morphologically unfamiliar agent. Dev Sci 7:425–430
Spelke ES, Phillips A, Woodward AL (1995) Infants' knowledge of object motion and human action. In: Sperber D, Premack D, Premack AJ (eds) Causal cognition: A multidisciplinary debate. Oxford Clarendon Press, pp 45–78
Striano T (2004) Direction of regard and the still-face effect in the first year: does intention matter? Child Dev 75:468–479
Towse JN, Redbond J, Houston-Price CMT, Cook S (2000) Understanding the dimensional change card sort: perspectives from task success and failure. Cognit Dev 15:347–365
Tronick E, Als H, Adamson L, Wise S, Brazelton T (1978) The infant's response to entrapment between contradictory messages in face-to-face interaction. J Am Acad Child Psychiatry 17:1–13
Tsuji A, Itakura S (2003) Attribution of dispositional states in infancy. In: Proceedings of Information Processing Society, Kansai Branch (in Japanese), pp 127–128
Woodward AL (1998) Infants selectively encode the goal object of an actor's reach. Cognition 69:1–34
Zelazo PD, Frye D, Rapus T (1996) An age-related dissociation between knowing rules and using them. Cognit Dev 11:37–63

Emergence of the Social Mind: Two Perspectives

Shoji Itakura[1]

The concept of the evolution of organisms is indisputable. All organisms, including humans, are products of evolution. Humans are related to other primates and have evolved from earlier hominid species over the past five to seven million years. Therefore, we can postulate that our minds are part of a wider evolutionary pattern discernible from the minds of nonhuman animals. Many psychologists acknowledge the fact that modern evolutionary theory is useful in explaining human behavior and cognition. However, Bjorklund and Pellegrini (2001) claimed that this theory has certain shortcomings from the developmental perspective. They pointed out three reasons for this.

> First, despite our confidence that humans are the product of a long evolutionary history, we are still unsure of the specific progression that took place from ape ancestor to modern *Homo sapiens*, as well as of the factors responsible for these changes. Second, although evolutionary psychology did not ignore development, it failed to treat it seriously in its theorizing. Third, although much had been written about psychological developmental issues from an evolutionary perspective, we found no overarching evolutionary perspective in developmental psychology.

This book intends to address that issue, particularly in the domain of social cognition. It is extremely important for ethologists and comparative psychologists to have a developmental perspective and for developmental psychologists to have an evolutionary perspective, based on Bjorklund and Pellegrini's (2001) conceptual background.

In this commentary, I would like to accomplish three things. First, I would like to summarize the relationship between ontogeny and phylogeny from a general perspective. Second, I would like to summarize the development of mentalizing in human children. Third, I would like to propose two new perspectives, which I believe are useful to reconsider the emergence of the social mind with respect to the developmental and evolutionary perspectives.

[1]Department of Psychology, Graduate School of Letters, Kyoto University, Yoshida-Honmachi, Sakyo-ku, Kyoto 606-8501, Japan

Phylogeny and Ontogeny

Actually, phylogeny and ontogeny are both forms of developmental changes in structure and function over time (Bjorklund and Pellegrini, 2001). Phylogeny refers to the changes in a species over a geological time scale. Phylogeny is used synonymously with evolution. Ontogeny refers to changes over the lifetime of an individual, and most of the time it refers to what is commonly known as development (Bjorklund and Pellegrini, 2001). Although both phylogeny and ontogeny refer to changes over time, there are a number of differences between the two. One of the most distinct differences is the time scale of the two types of development. Evolution occurs over a very long period of time, ranging from hundreds to thousands or even millions of years. We can infer evidence of evolution from fossil records. However, since there are no fossil records with respect to the behavior or mind of our ancestors, we also make inferences by examining behavioral, morphological, and genetic similarities and differences among extant species. In particular, we have termed *comparative cognitive science* as the domain of study dealing with the evolution of behavior and cognition in all species. As opposed to phylogeny, ontogeny occurs over a much shorter time scale. It is directly and constantly observable across different species, different cultures, and different historical periods.

Then, what do phylogeny and ontogeny have in common? Bjorklund and Pellegrini (2001) indicated the following with regard to this point: (1) the underlying mechanisms that stimulate changes in these two types of development are similar and (2) the processes affecting phylogeny and those affecting ontogeny are significantly similar. Bjorklund and Pellegrini concluded that human ancestors, whose ancient genes we carry, also developed from conception to adulthood, and perhaps insights from ontogeny will be able to aid us in better understanding phylogeny.

Development of Social Cognition

I will now outline the development of social cognition, particularly with respect to mentalizing (see Fig. 1).

The term "mentalizing," coined by Frith and Frith (2003), has the same implication as "theory of mind." Frith and Frith stated that "'theory of mind' was not to be taken literally, of course, and it certainly did not imply the possession of an explicit philosophical theory about the contents of the mind". They further stated that the theory of mind implicitly assumes that the behavior of others is determined by their desires, attitudes, and beliefs. These three elements are not states of the world but states of the mind. Although this consideration is crucial, I believe that mentalizing attributes mental states such as goal-directedness, intention, and the functioning of the mind to humans and nonhuman agents. In other words, mentalizing refers to the manner in which humans perceive nonhuman agents and attribute mental states to them. Thus, the development of mentalizing implies the development of a mind that discovers other minds (see Itakura, Chap. 9, this volume).

FIG. 1. Outline of the development of mentalizing.

I will now summarize some key transitions in the development of mentalizing based on Frith and Frith (2003):

a) Birth to 3 months: During this period, infants who are only a few weeks old smile and vocalize more to humans, and to human-like dolls, than they do to objects (Legerstee 1992). Eye movements and biological motion can attract infants' attention at an extremely early age. For example, infants track the movements of self-propelled objects (Crichton and Kuttner 1999). Three-month-olds also show more interest in the kinematic patterns of point-light displays of humans walking than in those of point-light displays of humans in random movement (Bertenthal et al. 1984). Infants of this age also show sensitivity to social contingency. Such sensitivity is the first step toward understanding people as social agents. Social contingency is a useful cue deployed by infants to distinguish themselves from others. During social interaction, an infant may recognize that people are interactive if he or she reacts to them contingently; therefore, sensitivity to social contingency is an important milestone during the development of social cognition.

b) Nine months onward: At this age, infants begin to engage in triadic interactions that involve the referential triangle of child, adult, and some outside entity with respect to which the child and adult share attention. Gergely et al. (1995) defined the ability of infants to reason with regard to goals, i.e., the "principle of rationality." Infants at this age can independently represent the goals of agents and the means used to reach them. The ability to represent goals and to reason in a rational manner are considered important prerequisites for the ability to represent intentions. Tomasello (1999) claimed that at around 9 to 12 months of age, infants begin to engage in interactions that are triadic in that they involve the referential triangle of child, adult, and object with respect to which the child and adult share their attention. Thus, infants at this age begin to flexibly and reliably look where someone is looking (gaze following), use adults as social reference points (social referencing), and act with regard to objects in the same way as adults do (imitative learning). In other words, infants begin to "tune in" to the attention and behavior of adults toward outside entities. Tomasello (1999) terms this drastic change of social ability the *nine-month revolution*.
c) Eighteen months onward: This developmental watershed, which marks the end of infancy, is significant for the onset of pretend play, which is considered an important precursor of the onset of theory of mind. Leslie (1987) postulates that a child of this age has to maintain representations of real events that are separate from the representations of thoughts that no longer need to refer to such events. Infants' reliable imitation of intentional actions performed by others, regardless of whether these actions help achieve their goal, is also observed at approximately 18 months, as demonstrated by Meltzoff (1995).
d) Two years onward: Children at this age begin to reliably grasp other people's desires. For instance, Repacholi and Gopnik (1997) observed infants' responses in a food-request procedure, to explore their understanding of other people's desires. They found that 18-month-olds were able to engage in some form of desire reasoning. This result suggests that children not only inferred that another person held a desire but also recognized how desires were related to emotions and understood something about the subjectivity of these desires.
e) Five years onward: Children at this age reliably understand false belief tasks that require the attribution of a false belief to others. However, understanding others' false beliefs is not the only goal of social development. After they master this, children begin to understand more difficult tasks that require the attribution of a belief about another person's beliefs, i.e., second-order tasks. At this age, children begin to understand metaphors, sarcasm, and white lies.

Finally, an implicit version of mentalizing, which emerges first and is concerned with desires, goals, and intentions, usually occurs around 18 months (Frith and Frith 2003). However, we believe that the ability to mentalize is based on aspects of social cognition in early infancy, such as sensitivity to social contingency, face recognition, gaze following, and biological motion. From the viewpoint of developmental cognitive neuroscience, clarifying the neural mechanism of mentalizing in early infancy is the next challenge.

Integration of Two Viewpoints

In this book, the contributors have attempted to integrate the perspectives of evolution and development in social cognition. However, I would like to include two new perspectives, which I am certain can be integrated with the two existing domains of evolution and development. One is *evolutionary developmental psychology*, the other being *developmental cognitive neuroscience*.

Evolutionary Developmental Psychology

Bjorklund and Pellegrini (2001) define evolutionary developmental psychology as the application of the basic principles of Darwinian evolution, particularly natural selection, to explain contemporary human development. I believe that this perspective can be applied to the contemporary development of any species. Bjorklund and Pellegrini further introduce four aspects in evolutionary developmental psychology, namely, how gene-environment interaction is interpreted, the role of domain-general mechanisms in explaining behavior, the significance of individual differences, and an examination of the role of behavior and development in evolution. In this book, Hirata (Chap. 2) refers to the mother–infant interaction in chimpanzees.

I also believe that this developmental trajectory must be examined in nonhuman species, and from this perspective I cite the study by Okamoto et al. (2002) as an example of a study that does this. Okamoto et al. investigated the development of the ability of longitudinal gaze-following in a chimpanzee by using a paradigm involving experimenter-given cues. One male chimpanzee was tested regularly from 6 to 13 months of age. There were four types of gestural cues provided by the experimenter, namely, tapping, pointing, head turning, and glancing. The chimpanzee fulfilled the criterion for the tapping condition at around 8 months of age; the pointing condition, at around 9 months of age; and the head turning condition, at around 10 months of age. By 13 months, the subject fulfilled the criterion for the glancing condition. The results suggested that the chimpanzee showed reliable gaze-following behavior in response to the experimenter's behavioral cues, including simple glancing, toward the end of 13 months. In addition, the authors reported that the chimpanzee's gaze-following behavior was controlled not only by the "social" properties of the experimenter-given cues but also by the stimulus-enhancing or local-enhancing peripheral properties. There were two conditions in the experiment: one was an incongruent pointing condition and the other was an incongruent head-turning condition. In the former condition, the experimenter gazed at and pointed to the target object with his index finger from the side of another object (called the distractor). In the latter condition, the experimenter oriented his head and eyes toward the target object from the side of the distracting object. The authors' hypothesis was that if the infant chimpanzee's responses were based on stimulus enhancement, it would choose the object closer to the experimenter (distractor). However, the results revealed that the subject looked significantly more often at the socially cued object in both the incongruent pointing and incongruent head-turning conditions. These results suggest that the infant chimpanzee employed the social cues presented by the experimenter. This is an extremely important finding.

Developmental Cognitive Neuroscience

Developmental cognitive neuroscience is the other viewpoint. Generally, not all measurements in psychology have such overt behavior as our target. The measures of underlying physiological processes can also be informative, particularly in infants and young children in whom overt behaviors are often limited. Recent years have witnessed some remarkable advances in the techniques for studying the brain, which enable researchers to examine not only the brain's anatomy but also its activity while people perform a variety of tasks. These techniques are applicable to social cognition or mentalizing. Today, new powerful methods and tools are available to cognitive neuroscience, which enables us to address, more directly than before, queries on mentalizing. A set of tools related to neuroimaging, the technique that generates "functional" maps of the brain activity, is based on physiological changes. Three current techniques are readily applied to development in normal children: event-related potential (ERP), functional magnetic resonance imaging (fMRI), and near-infrared spectroscopy (NIRS). These techniques are considered to be particularly useful in the case of infants and toddlers, in whom overt behaviors are often limited. There are a significant number of studies on the neural substrate of social cognition in adults (Adolphs 2006).

I will now introduce one of the typical study concerns, namely, gaze perception in infants. The direction of others' eye gaze is a crucial source of information in social interactions. Eye gaze also provides information about other people's communicative intentions and future behavior (Baron-Cohen 1995). Striano et al. investigated the functional relevance of gaze cueing in infancy (Reid and Striano 2005) and presented 4-month-old infants with videos of a face whose eye gaze was directed toward one of two objects. When exposed to both objects again, without the face, infants looked longer at the previously uncued object, indicating that they perceived it as more novel.

Based on this study, an important unaddressed question was raised: How do infants process the relation between another person and an external object? How do they use the information provided by an adult's eye gaze to guide their attention and process environmental information? Striano et al. employed an event-related brain potential (ERP) approach to explore this question

(Hoehl et al. 2007). This paradigm enables the direct investigation of the neural systems included in information processing even in the absence of overt behavior. Their study assessed how 4-month-old infants process the directedness of adult eye gaze in relation to objects in their field of view, which is the same as that of the face itself. They presented static photographs of faces whose eye gaze was averted to the left or right sides. One object was presented near the face, either on the same side as the direction of the eye gaze or on the other side. Their prediction was as follows: infants form a stronger memory representation for cued objects. This would be reflected by an enhanced positive slow wave (PSW), which is probably related to stimulus updating or encoding in 4- to 6-month-olds, during the observation of stimuli depicting the cued objects that are gazed at.

The results suggest that infants differentially process whether an adult's eye gaze is directed at one object or averted from it. The PSW at frontal sites was better when the eye gaze was directed toward the object rather than when it was directed away from it. They interpreted this finding as evidence that infants form a stronger memory representation for cued rather than uncued objects.

The wealth of knowledge that has recently been accumulated on social cognition in human children and nonhuman animals is extremely impressive. However, we still have a long way to go, and further empirical research is required to map the ontogenetic course and phylogenetic course with respect to social cognition.

References

Adolphs R (2006) How do we know the minds of others? Domain-specificity, simulation, and enactive social cognition. Brain Research 1079:25–35

Baron-Cohen S (1995) *Mindblindness: An essay on autism and theory of mind*. Cambridge, MA: MIT Press

Bertenthal BI, Proffitt DR, Cutting JE (1984) Infant sensitivity to figural coherence in biomechanical motions. Journal of Experimental Child Psychology 37:213–230

Bjorklund DF Pellegrini AD (2001) The origins of human nature. American Psychological Association, Washington, DC

Crichton M, Lange Kuttner C (1999) Animacy and propulsion in infancy: Tracking, waving and reaching to self-propelled and induced moving objects. Developmental Science 2:318–324

Frith U, Frith CD (2003) Development and neurophysiology of mentalizing. In: Frith C Wolpert D (eds), *Neuroscience of social cognition*. Oxford University Press

Gergely G, Nadasdy Z, Csibra G, Biro S (1995) Taking the intentional stance at 12 months of age. Cognition 56:165–193

Hoehl S, Reid V, Mooney J, Striano T (2007) What are you looking at? Infants' neural processing of an adult's object-directed eye gaze. Developmental Science 10:1–7

Legerstee M (1992) A review of the animate-inanimate distinction in infancy. Early Development and Parenting 1:59–67

Leslie AM (1987) Pretence and representation: The origins of "theory of mind". Psychological Review 94:412–426

Meltzoff AN (1995) Understanding the intentions of others: Re-enactment of intended acts by 18-month-old children. Developmental Psychology 31:838–850

Okamoto S, Tomonaga M, Ishii K, Kawai N, Tanaka M, Matsuzawa T (2002) An infant chimpanzee (*Pan troglodytes*) follows human gaze. Animal Cognition 5:107–114

Reid VM, Striano T (2005) Adult gaze influences infant attention and object processing: Implications for cognitive neuroscience. European Journal of Neuroscience 21: 1763–1766

Repacholi BM, Gopnik A (1997) Early reasoning about desires: Evidence from 14- and 18-month-olds. Developmental Psychology 33:12–21

Tomasello M (1999) Social cognition before the revolution. In: Rochat P (eds), *Early social cognition*. NJ: Lawrence Erlbaum Association

Subject Index